Thomas More Madden

The Principal Health-Resorts of Europe and Africa for the Treatment

of Chronic Diseases

Thomas More Madden

The Principal Health-Resorts of Europe and Africa for the Treatment of Chronic Diseases

ISBN/EAN: 9783337308841

Printed in Europe, USA, Canada, Australia, Japan

Cover: Foto ©berggeist007 / pixelio.de

More available books at **www.hansebooks.com**

THE

PRINCIPAL HEALTH-RESORTS

OF

EUROPE AND AFRICA

FOR THE TREATMENT OF CHRONIC DISEASES

BY

THOMAS MORE MADDEN, M.D., M.R.I.A.

VICE-PRESIDENT OF THE DUBLIN OBSTETRICAL SOCIETY
LATELY EXAMINER IN OBSTETRIC MEDICINE IN THE QUEEN'S UNIVERSITY IN IRELAND
PHYSICIAN TO ST JOSEPH'S HOSPITAL FOR SICK CHILDREN
EX-ASSISTANT PHYSICIAN TO THE ROTUNDA LYING-IN HOSPITAL, ETC., ETC., ETC.

PHILADELPHIA
LINDSAY AND BLAKISTON
1876.

TO

SIR MOSES MONTEFIORE, BART.,

AS A SLIGHT TRIBUTE OF ADMIRATION FOR HIS CHARACTER AS A
PHILANTHROPIST ;

HIS LONG-TRIED, UNTIRING, AND SIGNAL SERVICES TO HUMANITY
WHEREVER OPPRESSED AND SUFFERING ;

AND, ABOVE ALL, IN GRATEFUL REMEMBRANCE OF HIS PERSONAL
KINDNESS, IN FORMER YEARS AND DISTANT LANDS,
TO THOSE NEARLY AND DEARLY CONNECTED
WITH THE AUTHOR,

THIS VOLUME

IS RESPECTFULLY INSCRIBED.

PREFACE.

CHANGE of climate and mineral waters are each year more largely employed in the treatment of nearly every form of chronic disease. Notwithstanding the number of recent works on the subject, there is still a great deficiency of accurate information concerning even the most frequented health-resorts. For the majority of such books being either written by those residing in some sanatarium, the advantages of which are, perhaps unconsciously, exaggerated; or else being merely copied from these local advertisements, physicians at home are too often misled, and their patients abroad suffer the consequences. He who would attempt to supply the want of a comprehensive and reliable guide-book to the various foreign health-resorts should have extensive personal experience of many climes and places, and himself be a physician,—

> "—qui multorum providus urbes,
> Et mores hominum inspexit."

I may, therefore, premise that the following account of the principal Southern and Oriental winter asylums and Continental spas, resorted to by British valetudinarians, is the result of my own observations during several years of health-travel in many lands. However much my opinions concerning some of these places may be found to differ from those of other writers, at any rate they have not been formed hastily, nor without experience of the localities I have sought to describe. For instance, one chapter—that on Malaga—is founded on my observations there during three successive winters, and another—on Lisbon—was written after a still longer acquaintance with that climate.

For the favourable reception accorded to my former works on medical climatology and mineral waters I was chiefly indebted, next to the indulgence of my reviewers, to the circumstances which had caused me to visit or reside in the various health-resorts I described. The accuracy of my account of these places has been since attested by the freedom with which they have been appropriated by some recent authors. I therefore venture to hope that the following pages, which have been so thoroughly re-written and revised, as to form a substantially new work, may be found useful to the travelling invalid, as well as to his medical adviser, as a comprehensive and trustworthy hand-book of the foreign Health-resorts and Spas.

The therapeutic benefits derivable from change of climate and mineral springs have been somewhat unfairly ascribed solely to the cessation of drugging which generally takes place when a patient is sent abroad to a distant health-resort—

> " For change of air there's much to say,
> As Nature then has room to work her way;
> And doing nothing often has prevail'd
> When ten physicians have prescribed and fail'd."

Apart, however, from this negative service, the positive remedial influence of change of air, and the potent effects of mineral waters, have been recognized in every age, and were quite as familiar to ancient and mediæval medical writers as to those of the present day. Thus Hippocrates, throughout his works, and especially in the 45th of the second section of the ΑΦΟΡΙΣΜΟΙ, lays great stress on the use of change of air, of country, and of modes of life. Rhazes, the Arabian physician of the tenth century, enjoins patients suffering from chronic disease—" Mutare de loco in locum, suscipere itinera longa et indeterminata." Savonarola of Ferrara, the father of the famous monk, and himself perhaps the most re-nowned medical writer of the fifteenth century, dwells much in his treatise on nervous maladies on the advantages of change of air—"elongatio a patria." Gordonius, another authority of the same period, advises that, in the event of medicines failing in the treatment of dyspepsia or hypo-chondriasis, the patient should be sent into distant countries —"distrahatur ad longiniquas regiones." "Lælius à Fonte Eugubinus," says Burton, "that great doctor, at the conclusion of one of his consultations, most expressly approves of this above all other remedies whatsoever." "Many other things helped, but change of air was that which wrought the cure, and did most good." "The sum of all," adds Burton, in his "Anatomy of Melancholy," "is that variety of actions, objects, air, and places are excellent good in this infirmity, and in all others, are good for man and good for beast." Sir Thomas Brown, who conjoined the highest qualities of a physician with those of a philosopher, observes—"He is happily seated who lives in places where air, earth, and water promote not the infirmities of his weaker parts, or is early removed into regions that correct them. . . . Death hath not

b

any particular stars in heaven, but malevolent places on
earth which single out our infirmities and strike at our weaker
parts ; in which concern passengers and migrant birds have
the great advantages who are naturally constituted for distant
habitations, whom no seas nor places limit."

It is not enough, however, to make long journeys in pursuit
of health—

> " Ire per omnes
> Terrasque, tractusque maris, cœlumque profundus."

The invalid must make pleasant journeys, in good company,
and, above all, if possible, in good spirits. " I pity the man,"
says Sterne, " who can travel from Dan to Beersheba, and
cry ' 'Tis all barren ;' so it is, and so is all the world to him
who will not cultivate the fruits it offers." And how often
does the traveller meet in almost every place a tourist
of the family described by the author of the " Sentimental
Journey," who tells us that, " The learned Smellfungus
travelled from Boulogne to Paris, 'from Paris to Rome, and
so on ; but he set out with the spleen and jaundice, and every
object he passed by was discoloured or distorted. He wrote
an account of them, but it was nothing but the account of his
own miserable feelings. I met Smellfungus in the grand
portico of the Pantheon ; he was just coming out of it. ' It is
nothing but a huge cockpit,' said he. He had been flayed
alive, and divided, and used worse than St Bartholomew
at every stage he had come at. ' I'll tell it,' cried Smellfungus, ' to the world.' . . . ' You had better tell it,' said I,
' to your physician.' "

Very just was Yorick's advice ; for in ninety-nine cases out

of a hundred, this morbid state of mind arises either from
some physical disease, of which it may occasionally be the
only observable symptom, such as latent gout, hypochondriasis,
or derangement of digestion—or else, it is a precursor of
insanity, and, therefore, should be managed with extreme
forbearance and kindness, and treated by suitable remedies,
of which properly directed change of climate will often be
found one of the most efficacious.

The physiological influence of climate, although a subject
of the highest practical interest in connection with the therapeutic application of change of air, is far too wide a topic for
consideration in this place. That the human constitution is
powerfully acted on by climate is shown by the manifest
effects produced by change of country and of air on individuals in robust health, as exemplified by the history of all
colonization, which clearly proves that the various races of
mankind have each a certain fixed zone of temperature, within
the limits of which alone they can permanently thrive. And
though they may pass these boundaries for a short time, not
only with impunity, but, as will be shown in the following
pages, in many instances with the most striking remedial
results, yet if their exile from their native atmosphere be too
protracted, their constitutions become gradually altered, and
their health and vigour impaired. Moreover, should they now
marry and have children in the foreign climo, their offspring
will be in all respects a degenerate race.

Nearly all attempts on the part of Europeans to colonize
tropical countries have been attended with these results. Thus
the English, although two hundred years established in India,

have not succeeded to any extent in perpetuating their race in
that country, and would long since have been expelled from it
were not their numbers constantly recruited from the mother
country. The Portuguese settled before the English in Bengal
and Macao ; and not having been similarly reinforced, have
lost almost every trace of their European descent—

> " 'Twas not the sires of such as these,
>   Who dared the elements and pathless seas ;
>   Who made proud Asia's monarchs feel,
>   How weak their gold against Europe's steel ;
>   But beings of another mould,
>   Rough, hardy, vigorous, manly, bold ! "

The power of acclimation is, of course, greatest in the
inhabitants of the temperate zone, who, being habituated to
sudden and frequent atmospheric vicissitudes, as well as more
highly civilized than those of any other region, are better
enabled to protect themselves from the perturbations of the
atmosphere and inclemencies of the season. And it is within
this zone that human life attains its longest average duration.

The natives of either very cold or tropical countries affected
by their climate, and not highly civilized, are incapable of
extensive migrations, and succumb readily to great and
sudden change of temperature. Several years ago I had
an opportunity of observing the susceptibility of Australian
aborigines to change of climate in the case of some native
Western-Australian children, with whom I came to Europe.
With them the cold, damp air of this country disagreed,
and even the mildest atmosphere of Italy did not substitute
for that of their own country, nor ward off death, which
took place in one case within a couple of years time, not from

any distinct disease, but from general loss of health, apparently
the result not only of the alteration of the mode of life, but
still more the consequence of the change of clime, and, in
another was only averted by the return of the patient to his
native air.

Did space admit it would be easy to show that it is not
merely the physical constitution and bodily health of man
which are influenced by climate, but that his national peculiarities are to some extent affected by this cause.   Thus, for
instance, the cold and ungenial, but bracing, climates of Switzerland or Scotland, the rugged mountains they inhabit, and the
sterile soil they cultivate, have probably assisted in developing
the enterprising and hardy character of the Swiss mountaineer
or Scotch Highlander ; whilst the flat and monotonous land of
Holland, and its chill humid atmosphere, may thus have a part
in forming the sluggish disposition and phlegmatic temperament of the Dutch.   The frequent changes of weather and of
temperature to which the inhabitants of this country are exposed are, in like manner, reflected in the restless energy and
mental activity of the people.   Nor does the bright cloudless
sky, and unvaryingly warm and sunny atmosphere of the south
of Spain or Italy, less powerfully foster the poetic or imagina-
tive faculties—perhaps at the expense of the more practical
and sterner qualities of the mind.

If such be the influence of climate on entire races, modify-
ing not only their external form, the relative importance of the
functions of many of their internal organs and the diseases
that prevail in each isothermal zone, how great must be its
influence on individual men, more especially when these are

in a weak and delicate state of health, and therefore infinitely more susceptible of all the good effects of a suitable change of air, as well as of the injurious results of inhabiting an unhealthy or unsuitable climate.

The great majority of those who frequent the various winter resorts described in this work suffer from pulmonary disease, especially consumption or chronic bronchitis. In the first and second chapters I have shown how the former may, when, incipient, be warded off, or that, if already developed, in many instances it may be completely arrested in its progress, and also that the latter may be, in most cases, cured by a judicious change of climate. Besides pulmonary invalids a daily increasing proportion of the cases for which change of air is now prescribed is furnished by valetudinarians belonging to the wealthy classes of this country, who, having no necessary occupation, have become prematurely exhausted, or " used-up," by the laborious idleness of modern fashionable life, and imagine themselves to be ailing when, in fact, they are merely *blasé*. The complaints of such patients consist of disturbances of functions, and a long train of nervous symptoms attributable to want of occupation, weariness of life, and lassitude ; which, though they do not amount to actual illness, will, undoubtedly, if not controlled and counteracted in due time, merge into confirmed disease. Long experience has proved that the most effectual way to prevent such a result is to travel in pursuit of health, and thus create something to do, something to look forward to, and to live for.

If inaction of mind or body furnishes so large a number of valetudinarians requiring change of climate, the opposite condition is not less prolific of analogous cases. Amongst those whose position forces them to engage in too long-continued intellectual labour of any kind, whether it be in over application of particular faculties or in inordinate exertion to attain a status in science or in society; to avoid being trodden down or outstripped in the life-race of competition—there is no lack of ailing persons, dyspeptic and hypochondriacal, not absolutely sick, yet far from well, who must inevitably break down if they have not change of air, relaxation of mind, and removal from the sphere of their ordinary toils and cares.

The chronic diseases just referred to, in which change of climate is resorted to with such advantage, are by no means as numerous as those which may be cured by the judicious employment of some of the spas, which are fully described in the Second Part of this volume. Amongst the maladies in the treatment of which mineral waters are more efficacious than any other remedies, may be included gout and the protean complaints springing from the gouty diathesis, subacute rheumatism, rheumatic arthritis, scrofula in all its forms, dyspepsia and hypocondriasis, sterility, and most of the chronic diseases peculiar to women. In a word, morbid conditions the most opposite,—those occasioned by an inactive life and those originating from over-taxed powers of mind or body; plethora, and anæmia, the impoverished blood produced indirectly by imperfect digestion or resulting directly from hæmorrhage or disease, and the vitiated blood laden with excess of nutriment,—are all, as will be shown, in many instances curable by mineral springs. In addition to those who thus really require the special curative action of some mineral source, a large num-

ber now visit the health-resorts of the Upper-Engadine, the spas
of the Salzkammergut, or the other remote watering-places
described in this volume, who merely find in the state of their
health a plausible reason for indulging their innate and most
legitimate inclinations for travel. There is a fear of being
thought capable of seeking enjoyment for its own sake, peculiar to English professional men. But once introduce the plea
of delicate health, and talk of change of air and mineral waters,
and all such difficulties vanish. Though it certainly seems
strange why the over-worked clergyman; the lawyer, wearied
out by pouring over briefs; or the physician, jaded by what
Dr Johnson contemptuously described as — "a continual
interruption of pleasure; a melancholy attendance on
misery," should of all classes of society be reluctant to
admit that they go abroad for the pleasure as well as for
the moral and physical benefit of change of scene and change
of air.

Properly used, travelling is the highest pleasure of which
an educated and civilized man is capable, for it multiplies his
enjoyments by crowding a greater number of impressions and
sensations into a given space of time than anything else can
do. An old English voyager has left so excellent a picture
of the benefits of travel that I cannot refrain from quoting his
words—"But what," says Fynes Moryson, "if passengers
should come to a stately palace of a great king, were hee more
happy that is led only into the kitchen, and there hath a fat
messe of brewis presented to him, or rather hee who not only
dines at the king's table, but also with honour is conducted
through all the courts and chambers, to behold the stately

building, pretious furniture, vessells of gold, and heaps of treasure and jewells ? Now such and no other is the theatre of this world, in which the Almightie Maker hath manifested his unspeakable glory. He that sayles in the deepe, sees the wonders of God, and no lesse by land are these wonders daily presented to the eyes of the beholders, and since the admirable variety thereof represents to us the incomprehensible Majestie of God, no doubt we are the more happy, the more fully we contemplate the same."

To no tourist is it so important to know how to travel with advantage as to the valetudinarian pilgrim in pursuit of health ; for on this greatly neglected art depends the whole comfort and much of the benefit that may be derived from the journey. No people travel so widely as the English and none travel so well under difficult circumstances. Still no tourists seem to be less popular on the Continent. Nor is it improbable that those very qualities of endurance and resolution which so peculiarly fit British travellers for enterprises none other would undertake, incapacitate them from accommodating themselves sufficiently to foreign habits and customs. It has been well observed that "the traveller should consider himself a guest when abroad, and observe the same conformity with his host's manners, as he would were he paying a visit to a private house." But, unfortunately, this good advice is practically ignored by the majority of our travelling compatriots, who indulge in incessant depreciation of all the manners and institutions of foreign lands, or in invidious comparisons with something they think much better at home. As the Spaniards say, " Muchas veces la lengua corta la cabeza," and these rude speeches are among the principal causes of our unpopularity abroad.

It may seem superfluous even to allude to the respect due by the traveller to the religious observances of the countries he visits, however much they may differ from his own. But I have more than once been present in foreign cathedrals when tourists, in the outward semblance of gentlemen, and I must add some dressed as ladies, have, by their grossly irreverent and scoffing demeanour, wantonly outraged the most sacred feelings of congregations engaged in solemn worship. And, therefore, I feel justified in recommending any pilgrim in pursuit of health, whose zeal for his own faith is so strong that he cannot repress the manifestation of his intolerant contempt for the religious usages of others, not to seek a sanatarium beyond the limits of his own country.

Next to his prejudices, his luggage is the great incumbrance of the traveller, and like them it should be reduced to the minimum, for it is always far better to buy whatever one finds wanting abroad than to have the trouble of carrying about unnecessary articles.

The pilgrim in pursuit of health should, if possible, have some one of similar tastes and habits for his travelling companion. But the gregarious fashion of journeying now adopted by many is, I think, quite unsuitable for a valetudinarian, nor can I conceive any inducement which would persuade me to join one of the caravans of tourists who, not knowing anything of each other beforehand, bind themselves for a fixed period to perambulate foreign countries together under the direction

of a conductor, and may be daily seen scouring through
Continental cities after their leader, in the fashion of a flock
of sheep following the bell wether. But this is a mere matter
of taste and *chacun a son goût.*

With regard to the plan of this work I need not say much.
In the First Part I have given an account of the various
southern health-resorts, prefaced with some observations on
their special employment in different chronic maladies, particularly consumption and other pulmonary affections. In the
Second Part will be found a succinct description of the principal Continental spas, together with an account of the therapeutic effects of mineral waters in various diseases. And in
both parts I have endeavoured to combine all the general information on these subjects which may be useful to the prescribing physician with all those local details that are necessary to the invalid traveller. Bearing in mind, however, the
fact, too often lost sight of in similar works, that a hand-
book should be portable as well as accurate, in order to accom-
plish my purpose within the compass of a volume which will
not add unduly to the incumbrances of the travelling valetudinarian, I have condensed the descriptive portion as much as
was possible without omitting any matter that seemed to me
essential to its design. Should this book be fortunate enough
to secure the approval of those for whose use it is thus
intended, I trust on a future occasion to supplement it by a
similar volume containing a full account of all the health-
resorts and spas of Great Britain and Ireland.

Although, for reasons already alluded to, in the following
account of the foreign health-resorts I have relied mainly on

my own notes and recollections of the various localities described, I cannot omit acknowledging my obligations to the resident physicians in nearly all these places, to whom I am indebted, not only for valuable professional information and important local details, but also, in many instances, for much personal attention and kindness.

33 MERRION SQUARE, SOUTH,
DUBLIN, *July* 1876.

ERRATA.

Page 21, line 8, *for* "Tonas" *read* "Toros."
Page 30, line 18, *for* "Fandas" *read* "Fondas."
Page 139, line 22, *for* "Canards" *read* "Cornaro's."
Page 145, line 15, *for* "Homberg" *read* "Homburg."
Page 154, line 13, *for* "Homberg" *read* "Homburg."

CONTENTS.

PART FIRST.

ON CHANGE OF CLIMATE, AND THE SOUTHERN HEALTH RESORTS OF EUROPE AND AFRICA.

PART SECOND.

THE SPAS AND THEIR USES.

SOUTHERN HEALTH RESORTS
AND FOREIGN SPAS.

PART FIRST.

ON CHANGE OF CLIMATE AND THE SOUTHERN
HEALTH RESORTS OF EUROPE AND AFRICA.

CHAPTER I.

ON CHANGE OF CLIMATE IN THE TREATMENT OF PHTHISIS.

OF the large and yearly increasing number of invalids, more especially consumptive patients, who are now sent abroad to winter in the various southern health resorts described in the following pages, probably as many are injured by a wrong, or by a too tardy change of climate as are served by the judicious and timely use of this most valuable remedy. Patients as a rule will not abandon the comforts of their homes, or the avocations of their daily life, as long as they can cling to them ; nor do physicians generally prescribe this step until other treatment has failed, and then, when perhaps the disease is far advanced, the patient may be induced to try change of climate as a last resource.

The temperature of a locality, to which so much importance is popularly assigned, is no criterion of its climate as a health resort. Thus in Algiers and Malaga I have frequently seen a

A

sense of unbearable heat and exhaustion produced by the
" Simoom " and " Terral " winds, without any corresponding
alterations being occasioned in the readings of the thermometer. The fact being that invalids are painfully sensible of
variations in the hygrometric state and electrical condition
of the atmosphere, which are in no way indicated by this
instrument, so much relied on. The effects of temperature
depend chiefly on the hygrometric state of the air, and on
this is founded the division of winter resorts into dry, or
tonic, and humid, or sedative climates. A dry, warm atmosphere stimulates the action of the heart and lungs, especially
quickening the circulation in the vessels of the skin, and
increasing its secretion and that of the liver, and consequently
lessening the elimination from the kindeys. The diseases
which prevail in hot, dry, tonic climates are, febrile, nervous,
hepatic and cutaneous disorders. Most cases of chronic
bronchitis, chronic rheumatism, debilitating mucous discharges,
and some cases of asthma, are relieved in a very marked
manner by an equable, dry, and warm atmosphere. On the
other hand, consumption, attended by symptoms of much
irritation, with a short, hard cough; dry asthma, and some
cases of chronic bronchitis, where there is little expectoration,
would be injured by a tonic climate. In so called sedative
winter climates, the atmosphere is not only warm but humid,
or even, as in some places, saturated with vesicular moisture.
In such climates the eliminations from the skin and lungs are
diminished, and the air necessarily containing less oxygen
than an equal volume of dry air, the breathing is hurried, the
process of respiration is less perfectly accomplished, and the
nervous energy and muscular power of the emigrant invalid
are diminished. These sedative climates, however, are suitable
in some few instances of chronic pulmonary diseases, such as
spasmodic asthma and chronic bronchitis, accompanied with

great irritability of the pulmonary mucous membrane and frequent hard, dry cough.

A great change has come over medical opinion and practice with regard to the choice of health resorts for consumptive patients. Twenty years ago the great majority of these, irrespective of the type or stage of the disease, when sent abroad were recommended to visit sedative or humid warm climates, such as Madeira, Rome, or Pisa. At present, however, nine-tenths of such patients are sent to winter on the Riviera at Nice, Mentone, San Remo, or in Malaga, or Upper Egypt, all dry, warm tonic climates; and in the greater number of cases this preference for a dry climate for the phthisical is certainly well founded.

The mean temperature of a health resort is of much less importance for consumptive invalids than the rapidity and frequency of the transitions between its highest and lowest temperature, and preference should be given, *cæteris paribus*, to that place which possesses the most equable, rather than the warmest winter climate. In the same way it is not sufficient to know the annual rain-fall in any locality, but we should take into account the number of days on which it rains. Thus at Cannes the annual rain-fall is about five inches more than in London; but, notwithstanding this, Cannes is incomparably a drier climate than London; the number of rainy days in the former being fifty-two, whilst in the latter it amounts to one hundred and seventy-eight. There is no circumstance connected with any health resort for pulmonary invalids of greater importance than the force and direction of the prevailing winds, and yet how much this consideration is generally neglected, will be manifest to any one who examines the position of some of the favourite winter resorts of southern Europe.

Besides these, there are other reasons which serve to explain why change of climate so often fails to benefit phthisical patients. One is the fact, that invalids who have put themselves to the expense and inconvenience of going abroad for their health, are sometimes disposed to think that this should suffice for their cure; and, accordingly neglect all the other remedies, as well as those precautions which a consumptive patient requires in whatever climate he may reside. The last point I may allude to as a frequent source of fallacy respecting the value of change of climate in the treatment of consumption, arises from a faulty diagnosis by which other chronic pulmonary diseases are still confounded with it.

My experience as physician to three large institutions in which the diseases of children are brought under my care, has confirmed the observation made in my first work on climatology several years ago, that there is no class of patients in whom we may more confidently hope for the beneficial effects of change of air than in the case of, children predisposed by the scrofulous diathesis, or by hereditary taint to consumption. The climate chosen for the treatment of this predisposition to tubercular disease should be dry, bracing, and equable.

The division generally adopted of phthisis into three stages, is of practical utility in the remedial employment of change of climate. To be used with success change of air must be early had recourse to, and nothing can be more useless and more cruel than to send a patient in the last stage of consumption abroad, merely as a *dernier resort*.

In the first stage of the disease, a tonic, or dry, warm, maritime winter climate will be generally indicated. That is to say, that whilst the tubercular deposit and pneumonic inflammation is limited to a small portion of the pulmonary structure, the cough and dyspnœa slight, and the pulse not very quick, a change of air from this cold and humid climate, to a dry, warm atmosphere in winter, holds out a reasonable prospect of curing our patient. The climates which are best adapted for this condition are those of Upper and Middle Egypt, Western Australia, and Malaga.

In the second stage of phthisis, a more sedative winter climate is generally required than in the first. But even in this stage four patients may be sent to a tonic winter climate, such as Malaga, for one that may be recommended a sedative climate, such as Rome, Pisa, Madeira, or one of our home winter resorts of the same class, such as Queenstown, Torquay, or Sidmouth.

In the third stage of consumption, the expediency of change of climate becomes at times a subject of nearly as much anxiety to the conscientious physician as it is to the invalid under his care. In undertaking the risk of sending a patient in this stage of phthisis abroad, the physician must be guided altogether by the symptoms of each particular case, and the general condition of each patient ; and if the tubercular excavations be small, and the infiltration of tubercles and accompanying inflammatory action be limited to a small portion of the lung, the existence of the third stage of phthisis should not prevent us from giving our patient a chance of prolonging his life by a judicious change of climate. If, however, the patient's strength is much impaired; if the vomicæ are large or numerous ; if the pneumonic inflammation and the tubercular infiltration be widely diffused, and if profuse expectoration, colliquative sweating, and diarrhœa are rapidly accelerating the termination of the disease, then the only result of sending a person in this condition to a foreign sanitorium would be, to deprive him of the consolations and comforts of home in his last moments.

An eminent modern writer goes so far as to say—"If we

have come to the conclusion that a consumptive patient has tubercles we ought not to send him to Nice, Cairo, &c., but ought to let him live his last days amongst his own friends and die in his own house."* From my own experience, however, I can contradict this theory, as I know a considerable number of persons in whose lungs tubercles undoubtedly existed, and who, had they been treated in accordance with this doctrine, would have " died in their own houses " many years ago, but being treated by a judicious change of climate are now not only alive but in good health. I have myself attended patients who, when they landed in Malaga, could hardly walk from the jetty to their hotel on the adjoining Alameda, so weakened were they by hæmoptysis, unceasing cough, night sweats, and purulent expectoration, and who presented all the physical signs as well as the symptoms of a tubercular cavity in the lung; and yet within a few weeks I have seen the same invalids so improved that they could ride to a picnic, enjoy themselves there, and return home late at night apparently none the worse for the fatigue and excitement they had gone through. But I do not say that in such cases this improvement was often permanent, and though the patient with characteristic hopeful- ness might fondly persuade himself that the disease had yielded to the remedial influence of the climate, yet in the majority of instances the symptoms slowly returned, and the fatal event was only postponed, for some time.

Facility of access is a point of great importance in the selection of a climate for invalids. It is obvious that a patient should be sent to a place which may be reached by an easy journey, and that in case of need, he may communicate quickly with his physician and friends at home.

With few exceptions the localities resorted to in winter by

---

* " Clinical Lectures on Pulmonary Consumption," by Felix Von Niemeyer, M.D., p. 71.

consumptive patients are situated on the sea; and certainly this predilection is well founded, for such situations are more equable in temperature, being cooler in summer and warmer in winter, and less subject to sudden transitions or great extremes of temperature than inland places.

As a general rule, therefore, I would select a locality on the sea-shore for the residence of consumptive patients, not only for the reason just mentioned, but also on account of the facilities for reaching such places by sea. As far as my experience goes, there is no remedy so beneficial and so universally applicable in cases of incipient consumption as a sea voyage. And even in those cases of confirmed consumption, when circumstances render a long voyage, for instance to Australia, impracticable, the patient may generally travel to his destination more advantageously by sea than by land. It is evident that the monotony of a sea life, and the gentle unceasing motion of a vessel must occasion less excitement and fatigue than rapid land travelling would produce.

Change of air, however, appears to confer a sort of immunity on invalid travellers, who oftentimes when journeying undergo atmospheric changes and hardships which would probably have proved fatal to the same persons at home. Moreover, the change of living, of scene, and of place, and the freedom from the ordinary anxieties and duties of life, are as serviceable to the general health of a consumptive traveller, as the warm genial atmosphere of a southern winter resort is to the pulmonary complaint from which he suffers.

Preference should always be given to those winter resorts which present the greatest inducements and opportunities for open air exercise ; and no small part of the benefit derivable from removal to a southern health resort results from the opportunities afforded in the latter of being much in the open air, instead of being cooped up within doors as the patient

would be during the greater part of winter, had he remained
at home.

For those cases of phthisis in which the circumstances
of the patient do not admit of the expense, or in which his
state of health could not endure the fatigue of a journey to a
foreign sanitorium in winter, we sometimes very advantageously
have recourse to one of our moderately warm and equable
British winter resorts, such as Torquay, Hastings, Worthing,
or Bournemouth, which are all sedative, or slightly humid, as
well as sheltered climates. On the south-west coast of Ireland
there are several localities possessing similar climatic advantages, and more especially along the shores of Kerry and Cork,
which are exposed to the influence of the warm Gulf-stream,
and are well protected by mountain ranges from cold northerly
and easterly winds, such as Glengariffe and Dingle, where the
arbutus, myrtle, and other southern plants grow freely in the
open air all the year round, and which may be resorted to with
advantage in the same class of cases as Sidmouth or Torquay.
One of the most genial winter climates in the British Islands
is that of Queenstown.

It has been already observed, however, that these sedative
climates are by no means universally serviceable to phthisical patients, who more generally require a somewhat dry,
tonic, moderately warm, and bracing atmosphere. Of such
climates, unfortunately, the number is extremely limited
in these islands, being chiefly confined to a narrow strip,
about six miles in length, situated on the north-east coast
of the Isle of Wight, *i.e.*, the Undercliff, which undoubtedly
possesses the best tonic winter climate for consumptive
patients in the British Islands. There are, however, very
many instances in which change of air is urgently required
in the treatment of incipient phthisis, but in which the means
of the patient do not admit of residence sufficiently prolonged

in any of the health resorts described in the following pages. And in some of these cases the physician might be able to suggest a locality where the consumptive emigrant might find a climate suitable for the cure of his disease, as well as a fair field for the profitable exercise of his regained health and vigour. My own personal acquaintance with colonial climates is limited to Western Australia and the Cape of Good Hope. The former, although the least prosperous of the Australian colonies, possesses one of the best climates in the world, being peculiarly equable, moderately warm, dry, and bracing. The Cape Colony, on the contrary, is subject to sudden atmospheric vicissitudes, rapid changes of temperature, and violent storms, which must render it ineligible for this class of invalids. But the uplands of the adjoining colony of Natal appear to offer peculiar climatic advantages for phthisical immigrants.

It has been long observed that the inhabitants of elevated mountain districts appear to be peculiarly exempt from consumption, and an attempt has been made of late to turn this observation to practical account by recommending such localities as health resorts for the phthisical. It is, however, more than doubtful if the fact that the hardy mountaineers who inhabit Alpine districts and whose lives are passed under the most favourable hygienic conditions as regards pure air and exercise— the natural prophylactics against tubercular disease—are rarely attacked by consumption, can be regarded as a proof that these localities are therefore suitable winter resorts for patients already phthisical, and whose state of health would in such elevated, and oftentimes intensely cold and variable climates, probably confine them to the house in the new sanitoriums of the Engadine, or other Alpine resorts, during the greatest part of winter.

Still it would be difficult to deny or explain away the benefits which have been thus derived from mountain climates in the treatment of certain cases of phthisical disease or predisposition.

Some years ago the late Dr Archibald Smith * called attention to the exemption from phthisis enjoyed by the inhabitants of some of the highest districts of the Peruvian Andes, and the benefits derived by pulmonary invalids sent by the local physicians from Lima to certain Andine valleys, such as that of Jauja, upwards of 10,000 feet above the sea. Since then · similar advantages have been claimed by Dr Mattocks,† for the cold, dry, climate of the high table-land of the state of Minnesota, which lies nearly midway between the Atlantic and Pacific coasts of the United States. And also by Dr Weber,‡ Dr Küchenmeister,§ Dr Theodore Williams,¶ and other recent writers for certain more accessible European mountain climates, especially St Moritz and Samaden in the Upper Engadine, Davos in the Grisons, Köningswart near Marienbad, and Gorbersdorf in Silesia.

Of all the conditions necessary for a healthy climate one of the most essential is purity of the air; that is, freedom from malaria or miasmatic emanation, or from the atmospheric contaminations produced by the crowding together of vast bodies of men, or evolved by manufacturing or chemical processes. In this respect the Alpine climates of the Engadine

---

* Dr A. Smith " On the Climate of the Swiss Alps and Peruvian Andes Compared." " Dublin Quarterly Medical Journal," vol. 91, p. 339.

† Dr Mattocks' "Report on Minnesota." In Dr Horace Dobell's "Reports on the Progress of Practical and Scientific Medicine in Different Parts of the World," p. 8–24. London, 1870.

‡ "On the Treatment of Phthisis by Prolonged Residence in Elevated Regions," by Herman Weber, M.D., " Medico-Chirurgical Transactions," vol. lii. p. 223.

§ " Die Hochgelegenen Plateaus als Sanatorien für Schwindsüchtige," von Dr F. Küchenmeister, Vienna, 1868.

¶ Dr Theodore Williams " On the Effects of Warm Climates in Consumption." " Medical Chirurgical Transactions," vol. lv. p. 237.

or the Andine valleys of Peru are unquestionably better situated than any of the health resorts of the Mediterranean.

On the other hand, it must be borne in view that the height of a locality above the sea-level not only affects its temperature, which falls 1° for every 400 feet of altitude, but also, and still more directly, influences the pressure of the atmosphere, which at sea-level amounts to nearly fifteen pounds on every square inch of the surface of the body, and rapidly diminishes as we ascend above this ; and that the lessened atmospheric pressure of mountain health resorts must necessarily affect the balance of the circulation, giving rise to congestions or hæmorrhages, and would appear to be especially contraindicated in cases of consumption, already predisposed to hæmoptysis.

CHAPTER II.

ON CHANGE OF CLIMATE IN VARIOUS CHRONIC DISEASES.

THE therapeutic influence of change of air is by no means confined to the malady spoken of in the last chapter, but is equally applicable, although less generally recognised, in cases of chronic bronchial and laryngeal inflammation, asthma, hypochondriasis, and the incipient stage of mental disease resulting from overwork, chronic rheumatism, and rheumatic gout, and above all in hysterical affections, and some other of the chronic diseases peculiar to women, which will be again alluded to in a subsequent chapter.

There is hardly any chronic disease in which the remedial effects of change of air are so marked as chronic bronchitis, in cases of which I have often seen the mere change from town to a sheltered locality on the sea-side a few miles distant, cure a winter cough that no medicine could relieve. There are two varieties of chronic bronchitis requiring very different climates for their cure,—first, a form characterized by a hard, dry cough, with great irritability of the mucous membrane of the air passages; and, secondly, that in which the disease debilitates and saps the constitution of the patient by profuse and constant expectoration. In the first, Rome, Madeira, or any other of the sedative winter climates may be advised. The second, and by far the most numerous class of bronchitic patients require a tonic or dry winter climate, such as Western

Australia, Upper or Middle Egypt, Malaga, Nice, or Heyers. With some few patients Algiers, which holds an intermediate place between tonic and sedative climates, agrees better than any other. In chronic bronchial affections, however, it has been generally observed that occasional change of air is better than a continued residence in any climate however desirable.

The foregoing observations apply equally to the climatic treatment of chronic laryngeal and asthmatic affections.

Cases of chronic rheumatism, rheumatic-gout, and other diseases occasioned or aggravated by the sudden atmospheric changes, cold and damp, of this climate are necessarily benefited by a change to some warm, dry, and equable winter resort. I may here, however, caution rheumatic patients, who are sometimes recommended to winter in Pau, to avoid that climate; for not only is the disease in question very prevalent there, but also there is a great tendency to the sudden development of any latent cardiac affections which are so often met with in such cases.

In cases of dyspepsia, change of climate oftentimes proves an effectual remedy. I shall, however, reserve any observations on this subject for the second part of the present volume.

There are few patients so little benefited by physic, and so generally served by change of climate, as those who suffer from hypocondriasis, in which complaint the moral effects of travelling are no less marked then the physical. The primary action of change of air in such cases, however, consists in improvement of the digestive functions, soon followed by a diminution of nervous irritability. Constant travelling for a few months will generally be found more serviceable than a prolonged residence in any health resort. It matters little where the tour be made, provided always that it be in a dry and bracing atmosphere, the object being to obtain a complete change; and therefore the further from England and English associations the hypochondriac from these countries goes, the better.

"The saddest and most humiliating subject of thought," says Dr Johnson, "is the uncertain possession of the tenure of reason;" and, unfortunately, it is a subject daily brought more prominently before us, by the steady annual increase of late years in the number of the insane in this country. Any means, therefore, which affords even a hope of checking this disease in its incipient stage, deserves attentive consideration. And that we do possess such a means in change of climate is unquestionable.

There are now many predisposing causes of insanity to which our ancestors were much less exposed than we are. These evils are necessarily attendant on living in a densely populated state, where, by ill-directed education, divested of all moral control, the mind is prematurely exhausted and injured often at the very outset of the unceasing and increasing struggle not only for fame or fortune, but even for existence.

In no malady is the adage "prevention is better than cure" more applicable than in incipient insanity, and in no disease are the good effects of change of climate so obvious. By this measure not only is the patient for a time removed from the circumstances and cares of life, by which his mind was over-strained, whilst by the change of occupation and scene new and more wholesome thoughts are suggested.

> "Haply, the seas and countries different,
> With variable objects, shall expel
> This something—settled matter in his heart;
> Whereon his brain still beating, puts him thus
> From fashion of himself."

Conjoined to this, are the good effects of the moral restraint which such patients usually exercise over them-selves, when in the presence of strangers, before whom they often succeed in concealing the manifestations of those peculiarities and eccentricities, which, were they allowed to go on unchecked, would probably lead to confirmed mental disease. Dr Willis, whose treatment of George III. brought him patients from every part of Europe, remarked, that insane people who were sent to him for advice from the Continent more frequently recovered than his English patients did, and Esquirol, of Paris, makes a similar observation.

Some cases marked by excessive irritability require a mild sedative climate, and others, whose prominent symptoms are those of depression and languor, will demand a dry and stimulating atmosphere. But these varieties require a very careful appreciation and intimate knowledge of the symptoms of each case, and cannot be disposed of by any cursory general remarks.

In some of the diseases peculiar to advanced age change of climate will be found the best adjuvant to our efforts to "Husband out life's taper at its close." This is especially the case in climacteric disease, when the patient becomes conscious of a gradual decadence of all the mental as well as vital powers without any specific complaint, accompanied by longing for repose, which perhaps induced him to give up his long-accustomed avocations, and to retire to some quiet country place, in the futile hope that he may

> "Crown in shades like these
> A youth of labour with an age of ease!"

But very soon the individual discovers that long-accustomed habits cannot be suddenly abandoned with impunity, and that a man who for the best part of his life has been an actor on the busy stage of city life, cannot in his old age learn to interest himself in rural pursuits, and having leisure for reflection finds that inaction is not rest, and that an active mind having no employment will prey on itself; the result being a despondent spirit in an ailing body. Physic can do nothing to cure, and very little even to relieve such a state, and then it is that change of climate, which combines occupation with amusement, often proves an invaluable resource.

Invalid travellers should bear in mind, that southern health resorts are not necessarily places where the chances of life for the fixed inhabitants are any better than those of even the most unfavourably situated parts of our own cold, damp, climate. On the contrary, in many foreign winter resorts no system of hygiene exists, no effective sewerage is provided, and no sanitary laws are enforced; and it is always therefore essential to forewarn invalids going abroad to winter of this drawback, and to urge them in such places to select their residence in an open situation fully exposed to the sun and on a rising ground. In most southern climates there is a tendency to occasional sudden transitions from the prevailing genial temperature to a keen sharp atmosphere accompanying certain winds, which if they occur frequently, render the locality unsuitable for a health resort. Moreover, the houses in these places are generally constructed with the view of affording protection from the summer heat, rather than shelter from the inclemency of winter. The narrow streets which exclude the sun, the large fireless apartments, with uncarpeted floors, the imperfectly closing casements, the height of the rooms, all indicate the necessity of a cautious and guarded manner of living. Avoidance of the night air is essential for pulmonary invalids in warm southern climates, where there is generally not only a rapid abasement of temperature after sunset, but also a profuse fall of dew during the night.

I would further counsel every traveller in a warm climate to wear fine flannel inner clothing, no matter how hot the weather may be ; and also, as British prescriptions are seldom understood by foreign apothecaries, even when they may profess to dispense them,—a fact of which I have seen some unfortunate proofs whilst residing abroad,—I would recommend the invalid to find room amongst his *impedimenta* for a small stock of whatever medicine his physician may advise before leaving England.

It too often happens that when patients go abroad they think themselves released from any observance of medical rules, which are as necessary to the invalid in foreign health resorts as at home, and not only "throw physic to the dogs," but indulge, without restraint, the appetite which travelling seldom fails to bestow, even in southern climates, where the quantity of animal food and alcoholic stimulants that may be consumed in this country are no longer required, and will not be tolerated by the system, and where the invalid traveller should impress on his mind the old Salernitan precept—

" Si tibi deficiant medici, medici tibi fiant,
Hæc tria, mens læta, requies, moderata dieta."

CHAPTER III.

THE MEDITERRANEAN COAST OF SPAIN AND ITS CLIMATES.

THE health resorts of southern Spain are within easy reach of English patients by railway direct from Paris, or by sea from Marseilles, whence there are, almost daily, steamers for Barcelona, and thence on to Malaga, with stoppages at all the intervening ports. Having more than once tried each of these routes, I have no doubt that, for an invalid, the latter is the best.

Those who fear even the short passage across the stormy Gulf of Lyons may, as I have said, reach Barcelona by railway from Paris to Perpignan in twenty-five hours, thence by diligence to Gerona in eight hours, and again by train to Barcelona in four hours. My recollection of the fatigue and discomfort I endured during this journey is too vivid to allow me to recommend any ailing traveller to undertake it.

The inconveniences of land travelling in Spain, away from the lines of railway which now connect all the great cities with the southern French lines, render this mode of peregrination unsuitable for the majority of valetudinarians. The way-side nns or *ventas* are generally small, the usual accommodation being limited to a single apartment, which serves as kitchen and saloon. This is commonly crowded with muleteers, who pass a considerable portion of the night in chanting some interminable ballad to the tinkling accompaniment of a guitar.

In these hostelries the fare and lodging are still the same as it was in that venta of which Purchas, an English pilgrim who visited Spain in the fourteenth century, has left an account.

" Bedding there is nothing fair,
Many pilgrims it doth afaire ;
Tables use they none to eat,
But on the bare floor they make their seat."

Very soon, however, will such places be unknown, even in the bye-roads of Spain. Railroads are fast intersecting every province, the telegraph crosses every mountain, steamers visit each port; the *camino de atajo* will ere long be deserted ; the *arriero* and contrabandista metamorphosed into railway porters, and the " posada " will be a thing of the past.

To return from this digression on Spanish travelling, into which the diligence from Perpignan drove us,—Barcelona, which, from its position and importance, should be the capital of Spain, is situated on the sea, in a plain, nearly encircled by mountains, upwards of two hundred miles south-west of Marseilles, and contains a population of 150,000 inhabitants. The town is intersected by the *Rambla*, once the bed of a river, but now the most charming *paséo* in all Spain. Here are situated the principal hotels, of which the chief are the Fonda de Quatro Naciones and the Oriente, as well as most of the cafés and theatres. The streets of Barcelona resemble those of a French, much more than a Spanish town ; nor are the manners, dress, or even the language of the people Spanish, as most of the lower class speak only the Catlan patois.

The climate of Barcelona is humid and variable, being subject in winter and spring to cold northerly winds, which, alternating with a hot sun and warm southerly winds from the African coast, occasion an atmospheric constitution necessarily productive of pulmonary disease. The residents appear

to be well aware of the danger, since in the hottest summer
day and the coldest winter weather, they are to be seen constantly wrapped in the same heavy woollen cloak; and in
going round the hospitals I was struck by the number of cases
of phthisis and other forms of pulmonary disease.

Barcelona would be a very unfit winter abode for any consumptive or bronchitic patient. During the month of January
I have seen the thermometer at 7 A.M. as low as 36°, and the
highest temperature during the day was 59° at noon. The
cold winds which come down with great force from the high
mountains behind the town, the intense cold which occasionally occurs, and the rapid variations from a very high to a
very low temperature, all render this climate prejudicial to
consumptive cases.

The next port the steamer *en route* to Malaga touches at is
Valencia, which, could we credit some writers, should be that
happy clime that poets sing of, where—

> " Unbent by winds, unchilled by snows,
> Far from the winters of the west,
> By every breeze and season blest,"

the invalid might forget his ailments, and exult mentally and
corporeally in all the influences of a genial and balmy atmosphere and never varying sunshine. But that, unfortunately,
this is not the case a short view of the territorial and climatic
aspects of Valencia will show.

This province, of old a kingdom, extends about two hundred
miles along the Mediterranean, varying in breadth from thirty
to sixty miles, and mainly consists of a gradation of mountains
and slopes; which produce abundantly, figs, vines, and olives,
and are considered dry and healthy in summer, while the lowlands, which are principally converted into swamps for the
cultivation of rice, are justly regarded as the reverse.

The city of Valencia is about three miles distant from the

Mediterranean, one hundred and seventy miles to the south-west of Barcelona; and, together with its environs, contains a population of upwards of 100,000 inhabitants. The streets still retain a semi-Moorish character, being dark, narrow and tortuous; the houses, too, have the flat roof peculiar to the East, and are lofty and sombre-looking.

Valencia must be an extremely dull residence for foreign invalids; for when the Cathedral, Plaza de Tonos, and Hospital,—which latter is directed by sisters of charity, and is one of the most magnificent and best-conducted institutions for the sick poor that I have ever been through in any part of the world,—have been visited, little remains to be seen.

The climate of Valencia is dry but changeable, and subject to rapid variations of temperature, the thermometer after an intensely hot day falling to a very low degree immediately the sun has set, from which time the atmosphere becomes charged with dew. According to Dr de Minano, the mean temperature of the months in this city is as follows:—

January,	51°	July,	78°
February,	55	August,	77
March,	59	September,	73
April,	64	October,	66
May,	66	November.	59
June,	71	December,	46*

During the months of January and February my notes show that on three consecutive days the mercury was considerably below zero at 7 A.M., and I have noticed a difference of 15° between that hour and midday.

During the summer the heat is oppressive, the thermometer frequently standing for long periods at above 80° in the shade, nor is the sultriness of the atmosphere tempered by the sea-

* "Diccionario de Espana y Portugal," par El Doctor Don de T Minano, p. 174.

breeze, which in most other Mediterranean cities renders this bearable.

The prevailing winds are—the east, which passing over the sea is mild in winter, but at all seasons is humid; the west and south-west winds come next in frequency, and often blow with great force in winter, when they are cold and dry, and in summer are hot and parching. The north wind is also cold and dry, but is not prevalent, as the city is sheltered on that side by the distant mountains of Aragon. The most injurious, but, fortunately, the least frequent wind here is the south, which is moist and warm, and passing over the swampy rice grounds is pregnant with the seeds of malarious diseases.

A climate such as this can hardly be suitable as a health resort. That it is not one for the consumptive is unquestionable, and during my two visits to Valencia I ascertained that pulmonary diseases are prevalent.

Passing by Alicante and Cartagena, which in a sanatory point of view present no attraction to the traveller in pursuit of health, we next arrive at Almeria, the Portus Magnus of the Romans, now a very dull fourth-rate trading town, where most of the steamers stop a few hours. The view from the port is perfectly African in character. The barren coast and brown hills around, where not a particle of verdure grows, are in keeping with the ruined Moorish castle which stands on the summit of the mountain, and the small whitewashed houses below forming the town, the population of which amounts to about 19,000.

There is now nothing of interest to be seen in Almeria, which, however, was one of the chief cities of Spain under the Moors, and almost the last of their strongholds. The climate is nearly the same as that of Valencia, but the hardness of the drinking water renders it still more unsuited to invalids.

CHAPTER IV.

CADIZ, SEVILLE, AND GIBRALTAR.

THE south-west of Spain may be more easily reached by invalid travellers directly from England by sea than by any of the routes described in the last chapter. Every week there are departures from Southampton to Gibraltar, and also as often from Liverpool and London. The average duration of this voyage is five days, and its cost from Southampton £13. Many of these steamers touch at Cadiz, which struck me as one of the dullest towns in Spain,—the white walls which surround it hardly containing a single object of interest to a visitor, although this city is perhaps the most ancient in Europe, being founded, according to the local historians, three hundred years before Rome.

The climate of Cadiz affords a striking example of the common fallacy of regarding the physiological action of climate as in any degree indicated by the *mean temperature* of the locality. The mean temperature of Cadiz is 62°; winter, 52°; spring, 59°; summer, 70°; and autumn, 65°; and the average temperature of the months is—

January,	. .	51°	July,	. . .	70°
February,	. .	53	August,	. .	72
March,	. .	55	September,	. .	70
April,	. . .	59	October,	. .	67
May,	. . .	63	November,	. .	58
June,	. . .	68	December,	. .	53

The variations of temperature are here effected with extraordinary rapidity, the situation of the town exposing it to the full influence of the easterly winds, which sweep through the Straits of Gibraltar. During the winter land winds from a northerly direction are most frequent, and in spring the prevailing winds are those from the sea.

Catarrhal and bronchitic affections are common in Cadiz, occasioned by the atmospheric vicissitudes which render the climate unsuitable for any pulmonary patient.

From Cadiz to Seville is but the journey of a few hours, either by steamer up the Guadalquivir, or by railway *via* Puerta Santa Maria and Jerez.

By all true Spaniards Seville has always been considered the eighth wonder of the world. Their own proverb tells us—

> " Quien no la visto a Sevilla,
> No ha visto Maravilla."

The streets of Seville, as in all the Saracenic cities of Andalusia, are narrow and crooked; across them, in the hot weather, canvass awnings are spread from the over-hanging roofs of the opposite houses, and, owing to this contrivance, even in the intense heats of summer, when the neighbouring plain is burnt up to a barren waste, and when, as was the case during my visit, the Guadalquivir ran nearly dry, they maintained a delicious coolness. The principal streets are—the Calle de la Sierpe and the Calle Francos, both close to the Grand Plaza, which is one of the finest squares in Europe, and in summer is the fashionable lounge of the Sevillians, who here assemble to listen to the bands from eight or nine o'clock in the evening till long past midnight. The best hotels are in this square.

Of the many monuments of ancient art, Christian as well as Saracenic, which Seville possesses, there is none comparable to the Cathedral, which has no rival except St Peter's. Its foundation dates from the year 1407, when the chapter resolved to "erect a church so great and so good that there should be nothing equal to it." "*Fagamos*," said they, "*una Eglisia tal que la posteridad nos tengan por locos*" ("Let us build such a church that posterity shall take us to have been mad "). From that time, this body voluntarily gave up their revenues and lived in community, in rigid poverty for more than a hundred years, until, finally, the sacred edifice was completed in 1519. "I do not hesitate," says Mr Robertson "to characterise the Cathedral of Seville as the noblest temple in Christendom."

The Alcazar, which was built on the ruins of the palace of the Roman prætors of Seville by the western caliphs in the 10th century, is the most perfect specimen of Moorish art remaining in Andalusia, with the exception of the Alhambra of Granada.

Notwithstanding all its unrivalled attractions of an artistic and antiquarian character, Seville has little to recommend it as a séjour for the pilgrim in pursuit of health. The climate is characterised by the tendency to sudden variations of temperature, which is common to most of the southern cities of Spain, though in very few of them is this so marked as in Seville.

The winter is not as warm or equable here as it is on the sea coast. But even at this season the actual cold experienced is seldom very intense, for as Dr Gigot Suard states, "the average temperature during the coldest days is from 39° to 41° at sun-rise, and 50° to 55° during the rest of the day." *

In summer the excessive heat of the day contrasts strongly with the piercing cold of the nights. The cold wind, to which the sudden nocturnal abatement of temperature is due, brings

* "Des Climats sur le Rapport Hygiénique et Médical," p. 561.

with it, in addition to catarrhal and rheumatic affections, continued and intermittent fevers of a typhoid character, which are particularly severe and common in the environs of the city.

Such a climate, and the diseases to which it gives rise, for in winter catarrhs, bronchitis, and other pulmonary complaints, sometimes leading to consumption, are prevalent, should warn us that mistrusting the fallacious brilliancy of the sky, and purity of the atmosphere, we should beware of sending patients suffering from diseases of the respiratory organs, and especially phthisical invalids, to winter in Seville.

I have no doubt, however, that in certain other chronic maladies the climate of Seville might prove very beneficial. I would include under this head many cases of chronic gastric and liver disorders, and general relaxation of fibre, resulting from long residence in tropical climates, and which unfit the individual for withstanding the cold and damp of our northern winters.

The most southern point of Europe, Gibraltar, demands a brief notice in this place, having been recommended by some writers as a proper winter residence for invalids, and also because all the steamers from England to the South of Spain touch here. Gibraltar is situated at the entrance of the Mediterranean, fifteen hundred miles from Southampton, forming a rocky peninsula nearly three miles in length, and separated by a narrow sandy isthmus, which can be laid under water at any time, from the main land. The rock rises abruptly to a height of 1400 feet, and is intersected by deep gullies, which act as reservoirs for the rain water, and add to the general unhealthiness of the locality by the evaporation they cause in summer and autumn. The population of the town is upwards of 15,000, exclusive of the garrison, which varies from 4000 to 5000 men.

Excepting for military men Gibraltar is a most uninteresting

place. On no account should the invalid traveller be persuaded to see the "galleries" in the rock usually visited by
tourists, as in order to do so he must first undergo the long
ascent to the commencement of the cuttings, where he will
arrive fatigued and heated, and then suddenly will be exposed
to the bitterly cold, damp air, which rushes along these
passages, and which cannot fail to injure weakened or diseased
lungs.

A great drawback to Gibraltar is the want of good drinking
water, for that principally used is rain water, which being kept
in limestone tanks becomes a prolific cause of calculus and
other diseases.

The climate of Gibraltar is very similar to that of the
northern coast of Africa, to which its position so nearly
approximates, being, however, somewhat modified by the
almost insular situation of "The Rock." The mean annual
temperature is 64°, the maximum being 92° in July, and the
minimum 32° in February. The mean daily range of temperature is 13°. The mean temperature of winter is 58°; that of
spring 66°; summer 77°; and autumn 67°.

These thermometrical details cannot, however, be relied on
as indications of the climate, for we shall often find in Gibraltar
that when the sun is most powerful, and the thermometer
consequently stands highest, at the same time a cold, searching,
easterly wind prevails, and even in the warmest weather the
shady sides of the streets are often felt to be unpleasantly
chilly.

The position of Gibraltar between the Atlantic and Mediterranean accounts for the prevalence of strong winds, especially
from the east, on which side the town is much exposed.
Easterly winds occur on an average for 177 days annually, and
generally prevail from July to November, which is regarded as
the unhealthy season; for this wind, which is so violent as to

render the bay of Gibraltar unsafe for shipping while it lasts, being besides saturated with moisture, is no less injurious to the pulmonary invalid. The westerly winds, which prevail on an average for 188 days in the year, blow directly on the town, but are dry and clear, and are usually considered healthy.

The year may be practically divided at Gibraltar into two seasons, the rainy and the dry, the first commencing about the end of September with very violent rains, which fall at intervals until May, when the dry season succeeds. Thus in an average year, from the 4th of August to the 1st of November, only 5 inches of rain fell, while from the 1st of November to the 30th of January, inclusively, 29 inches fell. The average annual rain-fall, however, is but 34 inches.

The autumn months here are damp and unhealthy, the atmosphere is thick and foggy, and heavy dews fall at night.

The mortality at Gibraltar is, as it always has been, very high. "It is especially observable," says Captain Sayer, civil magistrate at Gibraltar, " that although the population has been gradually decreasing since 1840, the death-rate has been gradually increasing." The prevalent diseases are consumption, affections of the pulmonary organs, and fevers.

In the last Report of the Army Medical Department we find that out of a garrison of 4341 men, no less than 2543 were admitted into hospital during the year.*

From the preceding details it will be seen, that the climate of Gibraltar is prejudical to all invalids, and more especially so to consumptive patients. The combination of an intensely hot sun and cold wind, the great variation of temperature between the sun and shade, the badness of the water, and the evils peculiar to a small garrison town, all conduce to a state of things hurtful alike to the moral tone and physical condition of the inhabitants.

* " Army Medical Department Reports," vol. xv. p. 56, London, 1875.

CHAPTER V.

MALAGA.

MALAGA is one of the best winter-resorts in Europe for consumptive patients requiring a warm, dry, tonic climate. I
have had great reason to speak favourably of this place, not
only from my own personal experience of its benefits some
years ago during three winter seasons, but also from a considerable number of cases in which I have since then recommended this climate with advantage.

The city of Malaga is situated eighty miles to the eastward of
Gibraltar, on a deep and beautiful bay, surrounded by an extensive plain, the vegetation of which is almost tropical in
character, opening on the Mediterranean to the south, and protected on the north, west, and east, by the lofty mountains of
Ronda, Antequerra, and the Sierra Nevada.

From Paris this health resort may now be reached in three
days by railway *via* Madrid and Cordova ; but I should not
advise this very fatiguing journey to any invalid traveller who
may in preference select one of the routes mentioned in the
last chapter.

Of the early history of Malaga we have no authentic
record until the year 711 A.D., when the Moors invaded
Spain, from which time this place gradually rose into a
city of great commercial importance. Some vestiges of its
former greatness may yet be seen in the ruins of the castle on

the hill, and in the Atarazanas, or arsenal, in the falling walls
of which still rust the rings to which the Moorish galleys
were once secured, though it is now nearly half a mile from
the sea. The long-enduring sway of the Saracens in Malaga
terminated in 1487, when Ferdinard "El Catholico," after a
siege of three months, entered the town.

Since then, with the exception of its having been attacked
by Admiral Blake in 1658, and again captured by the French
under Sabastiani in 1810, Malaga has enjoyed a comparatively
tranquil existence down to a very recent period, when a series
of republican and communistic "pronunciamentos," or abortive
revolutions, having broken out and been put down, this place
has again settled down into a prosperous but dull commercial
town, and now contains a population of about 120,000
inhabitants.

The principal hotels are the Alameda, Victoria, and Oriente,
which are situated close together on the Alameda. The
usual charge in these "Fandas" is from thirty to forty reals,
or from six to nine shillings a day, this tariff including all
the ordinary expenses of living.

The Alameda or fashionable promenade, which at the commencement of this century was partly covered by the sea, is
now separated from it by a couple of intervening streets, and
extends from the port on one side to the Guadalmedina or
river of the city on the other, being about half a mile in length.
Owing to the gravelly absorbent nature of the soil, which dries
very quickly after rain, the Alameda affords the best promenade in Malaga, unless when the *terral* or *levante* winds,
to which it is exposed, prevail. At all other times this walk
is the great resort of the foreign visitors. Here, too, may be
seen the sturdy contrabandista from Ronda, wrapped up to the
eyes in the folds of his ample cloak, basking in the warm
sunshine, or the tall peasant from the mountains in his pictur-

esque Andalusian costume, slowly stalking along with all the
grave dignity of an ancient Roman.

> " There of Numantian fire a swarthy spark
> Still lightens in the sun-burnt native's eye ;
> The stately port, slow step, and visage dark,
> Still mark enduring pride and constancy."

Next, perhaps, passes a solemn Padre, reciting his breviary
beneath the shade of a *sombrero* nearly three feet long, or more
numerously the dark-eyed *señoritas* gracefully trip along,
exchanging electric glances and telegraphic signals from the
pliant fan with the young *caballeros*, who, during the time
of promenading, ride round and round the Alameda.

All the public buildings of Malaga are small and tasteless
except the Cathedral, a vast structure commenced in 1528, and
which still remains unfinished. The choir contains some very
beautiful specimens of wood-carving, attributed to Alonzo
Cano, as well as a few good pictures. These, however, can
hardly be seen, owing to the Spanish custom of darkening the
windows, although the general effect, is certainly rendered
more impressive by—

> " The solemn gloom,
> Of the long Gothic aisle and stone ribb'd roof,
> O'er canoping shrine and gorgeous tomb,
> Carved screen, and altar glimmering far aloof,
> And blending with the shade."

There is a great contrast between the modern part of
the town near the Alameda, and the older portions, which
have undergone little change for the last three centuries. The
latter are connected with the port by a street which intersects
the city, crossing the Plaza de la Constitucion and the Plaza
de la Merced, and terminating in the Calle de la Victoria, so
named from the triumpha entry of Ferdinand and Isabella
into Malaga. The streets in this quarter are extremely narrow

and dark, and the massive houses, with their grated windows, have externally a gloomy and eastern appearance.

The rooms are spacious and lofty, though for the most part they are but sparingly furnished. Fire-places are generally unknown here, their place being supplied by the *braseros* or pans of charcoal, which are but a poor substitute for the cheerful blaze of an open fire; and the invalid on those few occasions, when he may feel inclined to regret the comforts of an English fire-side, will do well to wrap himself in his cloak, or retire to bed, rather than sit poisoning himself over the noxious fumes of smouldering charcoal.

With respect to the living, it must be admitted that the meat in Malaga is very inferior, in every respect, to that used in this country. The fish, however, is so good, and of so many various kinds, as in great measure to make up for the shortcomings of the animal food, and invalids may manage to live very well and very cheaply at the hotels on fish, kid, poultry, game, especially partridge, which is a standing dish here, even if they do not choose to venture on the meats served at the table d'hôte.

Spanish cookery is generally considered as intolerable by British travellers, but I think this is mere prejudice; and that garlic and oil, which enter so largely into all culinary operations in Spain, are absolutely necessary (moderately used) to supply the want of fat and of flavour in the meat, and to render it digestible.

The physician who sends his patient to Malaga, should impress on him the great importance of choosing an apartment having a southern aspect, as there is often a difference of 10° in the temperature of rooms facing the Alameda, and those at the back of the hotels. Besides the mere warmth, the front rooms are more cheerful, and enjoy the advantage of free exposure to the light and sun, a very essential matter for a

pulmonary invalid, who should recollect the Italian proverb—
" Where the sun does not enter the doctor must."

The hygienic condition of Malaga is as defective as it can
well be. In a great many of the houses there is no provision
for sewerage of any kind, and even in the more civilized part
of the city, on the Alameda, the drainage is very bad indeed.

It might be anticipated, from the great antiquity of this city,
and the high civilization of its successive conquerors—Greeks,
Romans, and Saracens—that the antiquary might here reap
an abundant harvest of treasures of the olden time, but such is
not the case. The various races who have in succession peopled
this fair land, have all, with envious haste, endeavoured to
obliterate any trace of their predecessors. For instance, the
beautiful Moorish arch of the Atarazanas, which time had so
long spared, was not long since built up with brickwork and
totally defaced ; and of the Greek and Roman monuments
described in the " Conversaciones Malaguenas," at the end of
last century, hardly one now remains.

The chief resource for invalids in Malaga lies in the beauty
and variety of the walks and rides in the vicinity, which,
thanks to the usual fine weather, are nearly always accessible.
One of the most beautiful of these excursions is that along the
Granada road, behind the town, between the cemetery and
Moorish aqueduct, and ascending the mountain until we come
to the point where the road turns away from the sea, the view
from which is one of the finest imaginable. In front rise the
snowy peaks of the Sierra Nevada ; to the south, the Atlas
Mountains of the opposite African coast, fully eighty miles
distant, are clearly defined ; while from west to east the vista
includes an uninterrupted prospect of the Mediterranean from
Gibraltar to the point near Velez Malaga. From this vast
range of view, the eye falls back with relief on the fertile plain
around the town, thickly planted with groves of oranges, olives,

C

and sugar-cane, and every eminence covered by the vine. In such a scene the invalid may, for the moment, forget all the ills that flesh is heir to, and abandon himself to the enjoyment of—

> " Forest and village, lawn and field,
> Ocean and earth, with all they yield
> Of glorious or of fair."

Another favourite ride is through the Vega, by the Churriana road to the grounds of " El Retiro," the country seat of the Conde of Alcolea, some five miles from the town. These gardens, when I first visited Malaga, were much resorted to by pic-nic parties, and I have here seen invalids, who at home never breathed the noon-day air except through a respirator, without any precaution whatever, dancing in the open air, in December or January, and suffer from no consequent ill result. Very pleasant excursions may also be made to Velez Malaga, Alhaurin, and Antequera, and to the sulphurous baths of Carratraca, which are much resorted to in cutaneous affections, as well as in some cases of chronic rheumatism and dyspepsia, by those who can bear with impunity the attendant fatigue and roughing; and now by railway to Cordova and Granada.

Society in Malaga is almost entirely confined to the mercantile class, many of whom are eminently hospitable and courteous to strangers ; among the ladies, too, there are not a few who fully justify the proverb that " Las Malaguenas," *son muy halaguenas, "* are very bewitching." The Malaguenians of the lower order, it must be admitted, bear a very indifferent character throughout the rest of Spain, being regarded as addicted to gambling, fond of drinking, not recognising the distinction between *meum* and *tuum*, and especially are given to the use of the knife, which even in trivial altercations is drawn as readily as the fist would be resorted to in England.

Such are the characteristics which distinguish the lower order of Malaguenians from Spaniards generally, who, and more

especially the Northern Spaniards, are as kindly, brave, trustworthy, industrious, and self-reliant a people as any in
existence. But in Malaga the great deterioration is probably
in some degree occasioned by the exciting and often variable
nature of the climate reflected in the passionate and uncertain
character of the inhabitants, who are further altered from the
national type by their semi-Moorish descent, and by the
admixture of races common to all large seaports.

The climate of Malaga is of a dry, warm, and equable
character, the thermometer varying little during the day, except
when the "terral," or "levante" winds prevail, both of which
produce great and rapid changes in the temperature. With
this exception the mean daily variation is very slight, amounting according to my observations to about 3° per diem during
the winter months. Immediately after sunset, however, there
is a very sudden fall of temperature, accompanied, especially
during certain winds, by so profuse a fall of dew as to render
it unsafe for the invalid to venture out of doors at this time.

The mean annual temperature of Malaga is 65°, or 15° higher
than London, 1° lower than Algiers, 9° higher than Pau, and
7° lower than Cairo. The mean temperature of winter is 55°,
or 16° higher than London, exactly the same as Algiers, 13°
higher than Pau, and 3° lower than Cairo. The mean temperature of spring is 68°, or 20° higher than London, 2° higher than
Algiers, 14° higher than Pau, and 5° lower than Cairo. The
mean temperature of summer is 78°, or 16° higher than London,
1° higher than Algiers, 8° higher than Pau, and 7° lower than
Cairo. And lastly, in autumn the mean temperature of Malaga
is 60°, or 9° higher than that of London, 2° lower than Algiers,
2° higher than Pau, and 11° lower than that of Cairo.

I was indebted to the late Dr Shortcliffe of Malaga for the
following valuable table, showing the—

Average Temperature at Malaga for a period of Ten consecutive Years.

Months.	Mean Temperature.			Highest Temperature.			Lowest Temperature.		
	8 a.m.	2 p.m.	11 p.m.	8 p.m.	2 p.m.	11 p.m.	8 a.m.	2 p.m.	1 p.m.
January..	53·8	57·5	53·1	59	62	57	50	53	48
February .	54·6	57·3	54·9	60	64	61	46	52	44
March ...	57·8	60·1	59·1	64	68	65	54	57	53
April	61·4	64·2	60·9	66	70	67	58	58	57
May	66·1	69·1	64·8	72	76	70	60	68	61
June	72·6	78·1	73·8	76	79	78	71	71	70
July	76·8	79·9	76·1	80	85	79	75	77	73
August ...	77·6	79·9	76·9	80	86	81	74	78	75
September	72·5	76·1	73·8	79	84	77	69	75	69
October...	66·1	70·1	67·5	74	76	74	60	63	60
November	59·6	62·5	59·1	66	67	65	48	54	46
December	54·8	58·5	55·1	61	64	63	44	52	46

According to my own observations during the winter, the mean temperature of the month of December was 59½°; this is calculated upon observations taken at 10 A.M. and 3 P.M. each day, and therefore only shows the temperature of the day. The mean daily variation between these hours was 2½°, the greatest variation was 5°, and the smallest variation 2°. The highest temperature observed was 68°, and the lowest was 52°. In January the mean temperature of the month was 57°: the highest 61°, and the lowest in the day 50°. The greatest variation in the day was 9°, the smallest 2°, and the mean daily variation was 3°. During February the mean temperature was 58°, the highest temperature 66°; the lowest 50°; the greatest variation in the day 7°, the smallest 2°, and the mean daily variation 3½°.

The annual rain-fall is comparatively small, and the month of February is the most rainy period of the year. During the three occasions I passed that month in Malaga, the weather was generally so wet and gloomy that invalids were confined to their rooms for two-thirds of the time. This rain is usually of a tropical character, falling in large drops and with much force. When February has passed, the quantity of rain that falls during the rest of the year is very small, amounting on an average to 16 inches 5 lines per annum. The total number of rainy days observed by Martinez during nine years was 262, or about 29 days annually, which is 90 days less rain than in Pau, 88 less than Rome, 44 less than Madeira, 41 less than Algiers, 45 less than Nice, and 15 more than at Cairo.

The mountains beyond the Vega almost completely shelter the town from the north and west winds. The prevailing winds are the east, or *levante*, which is cold and humid in winter, the north-west, and the south-east.

The north-west wind, or " Terral," is termed *un viento fatal*. It rushes through a gap in the Antequera Mountains, along the valley of the Guadalmedina, and arrives in Malaga laden with fine sand, which irritates the pulmonary mucous membrane. In summer this wind is hot and dry, giving rise to a highly irritable state of mind and body. In winter the Terral, sweeping down the snow-clad mountains, is intensely cold as well as dry. Its ill effects, however, cannot be measured by the thermometer alone ; for the force and rapidity of its motion, its aridity, and the quantity of impalpable sand it suspends, all combine with its low temperature to injure the valetudinarian visitor, who should therefore be forewarned to remain within doors as long as this wind prevails, which fortunately is seldom more than three consecutive days.

The connection between defective hygiene and epidemic disease has been too well illustrated in Malaga, which has been repeatedly devastated by plague, yellow fever, cholera, and other zymotics. In the first edition of my work " On Change of Climate," I gave a very detailed account from old records of no less than twenty-two pestilences, which have almost depopulated Malaga at different times between 1493 and

1804. The earlier of these appear to have been epidemics of genuine oriental plague, while the latter generally assumed the form of yellow fever. Of late years these pestilences have not returned, but Asiatic cholera has proved very fatal on several occasions.

The prevailing diseases here are fevers, especially the ex-anthemata, intermittent, and bilious fevers, acute diseases of the air passages, gastric affections, purulent and strumous ophthalmia, and elephantiasis arabum, elsewhere so rare in Europe, but of which I have seen well-marked instances in Malaga, where in one year no less than seven cases of this disease were admitted into the civil hospital.

Senor Martinez Y. Montes, Physician-in-Chief to the Military Hospital, has collected tables from which we find, that in January the number of deaths was larger than in any other month, while in May it was smaller. The total number of adult deaths during the year was 9049. The greatest mortality from any one disease was 711 from dropsy, and the smallest was one death from hydrophobia. Acute and chronic diseases of the respiratory organs, not including consumption, were the cause of 1208 deaths. The mortality from phthisis was 407, more than half of which, or 234, occurred in the civil hospital. Cerebral affections are apparently a prolific source of mortality here; 742 fatal cases of these complaints occurred, of which no less than 407 are set down to apoplexy.*

In Malaga the deaths occasioned by consumption among the native population are less numerous than in any other European southern locality resorted to by invalids in winter. This is, I think, a very significant fact; for, as we might naturally consider, a place where the mortality from phthisis was very great an unsuitable residence for phthisical invalids, so a city like Malaga, where the mortality from that disease is remark-

* Martinez, "Topografia Medica de la Ciudad de Malaga," p. 499.

ably small, should be a favourable locality for such patients. Its superiority in this respect to our own climate is sufficiently proved by the fact, that in this country 125 deaths out of every thousand are caused by consumption, while in Malaga only 34 deaths per thousand result from that disease.

When, however, we consider the mortality from all chronic diseases of the air-passages, we find the superiority of this climate is not so conspicuous as it might at first appear; for the deaths in Malaga from all chronic affections of the respiratory organs are very nearly as great as in London, amounting, according to Senor Martinez Y. Montes, to nearly one-ninth of the entire number recorded.

It is a remarkable fact that of late years, ever since this town has acquired a renown as a winter residence for phthisical patients, the mortality from this disease has increased notably amongst the native population. And thus a belief in the contagious nature of phthisis has now become so general in Malaga, that many Spanish lodging-house proprietors refuse to admit phthisical invalids as inmates. This opinion appears to have more foundation in fact than is commonly supposed, and I cannot doubt that, in hot climates at least, constant communication with consumptive patients, and especially sleeping in the same room with them, is very likely to prove injurious to persons in delicate health, and may determine the occurrence of phthisis in individuals who might otherwise have escaped this malady.

The diseases which I have seen · most benefited by the climate of Malaga were, consumption in its first stage and the cachectic state which immediately precedes the deposition of tubercles in the lungs. These cases may sometimes be frequently cured, and more commonly the progress of the disease may be checked for some time by the action of the climate. Cases of chronic bronchitis and humoral asthma in

elderly persons occasionally improve materially here; in other
instances, however, of a more irritative form of either complaint,
the climate of Malaga is too tonic or exciting, and may produce
much injury if improperly resorted to.

Another class of patients on whom this climate may be
expected to exert its most favourable influence, are children
predisposed to, or suffering from, any form of scrofula, whether
manifesting itself externally by swelling of the lymphatic
glands of the neck or elsewhere, or seated in the mesentery, or
assuming those premonitory symptoms which warn us that
tubercular disease of the lungs is not far off, and that nothing
is wanting but some slight exciting cause to call consumption
into active existence. In such cases a short residence in the
dry and tonic atmosphere of Malaga will oftentimes work
wonders.

It is quite as important to know what patients should avoid
any climate as to know what class should select it. Now, considering the great mortality from cerebral affections, and particularly apoplexy, in Malaga, I think that those predisposed to
such diseases should not choose this town for their residence.
Nor would I send a patient suffering from chronic rheumatism,
or rheumatic arthritis, or neuralgia, to Malaga; as the great
difference between the temperature of the day and night, and
the heavy dews that fall after sunset, render this town in such
cases inferior to other climates, such, for instance, as Western
Australia, Upper Egypt, and in some cases even Nice. Nor
can Malaga be advised to dyspeptic and hypochondriacal
invalids, as the dietary there is not generally suitable for these
cases.

The climate of Malaga was considered by the older native
writers, Don Fernandez Barea, Padre Garcia de la Lena, and
others to exert a relaxing and unfavourable influence on the constitution of young persons under the age of puberty. But they

regarded this locality as an advantageous residence for the old, and thought the climate a propitious one for them, and calculated, by aiding in the alleviation of the many physical annoyances of the aged, to add to the span of their existence.

44

the churches, defaced the paintings, and sold for waste paper
most of the valuable libraries of the conventual establishments.
The only one which escaped, and still exists, is the Irish
Dominican convent, which is some 200 years old.

Of late years, since I first visited Lisbon, the sanitary condition of the city has been much improved. The streets are
cleaner, the sewerage is better, and in all the principal thorough-
fares gas has superseded the dim oil lamps that formerly gave
a faint twinkle through the night. But in the upper and older
quarters ancient customs still prevail. On my last visit to
Lisbon I was practically reminded that the old cry of *agoa vai*,
"water beware," though not so commonly heard as it was, is
not yet a thing of the past. On this occasion, returning home
one dark night, I heard the cry, and had barely time to take
refuge in an open doorway before the unsavoury shower fell
close beside me.

A very short stay will suffice to visit all the sights of Lisbon,
but an almost inexhaustible fund of beauty and variety will be
found in the various excursions in the adjacent country; and
foremost among these is that to where

"Cintra's glorious Eden intervenes,
In variegated maze of mount and glen."

Cintra is now within an hour's drive from Lisbon by the steam
tramway. It would be utterly impossible, in the narrow limits
at my disposal, to attempt any description of the beauties of
this place, which in summer affords a delightful retreat from the
intense heat of Lisbon. For most invalids, however, the transition would be far too great. Thus, in the middle of June, I
have found the temperature of Cintra nearly 20° lower than
that of Lisbon.

The climate of Lisbon is warm, but humid and variable.
The mean annual temperature is about 61°; the mean temperature of winter is 54°; spring, 59°; summer, 68°; and of autumn,

LISBON.   45

59°. The minimum temperature of the year is seldom under
35°, but occasionally a slight frost occurs during the night in
December. The maximum annual heat in the sun is towards
the end of August, when the mercury at 2 P.M. sometimes rises
to 133°. The extremes of heat and cold annually observed
differ by between 50° and 60°.

These thermometrical details are deduced from the scattered
observations of various native and foreign observers, from
some tables quoted by Mr Murphy, and especially from two
papers by Senhor Franzini, in the Transactions of the Royal
Academy of Lisbon,* all of which I have tabulated together.

No sufficient register of the hygrometric state of the atmosphere has been kept, but it is well known by every observer
that the air in winter is here laden with vesicular moisture.
Senhor Franzini states that humid winds predominate over dry
ones in the proportion of 199 to 165.

Rain usually falls in heavy showers of brief duration. Thus
the number of days of settled rain in the year are comparatively
few, not exceeding 63, while the amount of rain is considerable,
amounting to upwards of 30 inches annually. One-half of this
quantity falls in winter, from November to the middle of
February; the remainder being nearly equally divided between
spring and autumn, as there is little or no rain from June to
October. During the rainy weather the south-west wind is
prevalent.

To the stranger recently landed in Lisbon from England, in
winter, the weather will at first appear by comparison mild
and genial; and so indeed it very often is. But still at this
season cold and damp winds from the westerly points are
prevalent. And although frost is a rare phenomena in the

* "Observaçoes Metèorologicas Feitas na Cidade de Lisboa, &c., par Marino
Miguel Franzini, in Memorias da Acadamia Real Das Sciencias de Lisboa,
tom. v. p. 92, 125, et tom. vii. p. 61, ad 93.

the churches, defaced the paintings, and sold for waste paper most of the valuable libraries of the conventual establishments. The only one which escaped, and still exists, is the Irish Dominican convent, which is some 200 years old.

Of late years, since I first visited Lisbon, the sanitary condition of the city has been much improved. The streets are cleaner, the sewerage is better, and in all the principal thoroughfares gas has superseded the dim oil lamps that formerly gave a faint twinkle through the night. But in the upper and older quarters ancient customs still prevail. On my last visit to Lisbon I was practically reminded that the old cry of *agoa vai*, " water beware," though not so commonly heard as it was, is not yet a thing of the past. On this occasion, returning home one dark night, I heard the cry, and had barely time to take refuge in an open doorway before the unsavoury shower fell close beside me.

A very short stay will suffice to visit all the sights of Lisbon, but an almost inexhaustible fund of beauty and variety will be found in the various excursions in the adjacent country; and foremost among these is that to where

" Cintra's glorious Eden intervenes,
In variegated maze of mount and glen."

Cintra is now within an hour's drive from Lisbon by the steam tramway. It would be utterly impossible, in the narrow limits at my disposal, to attempt any description of the beauties of this place, which in summer affords a delightful retreat from the intense heat of Lisbon. For most invalids, however, the transition would be far too great. Thus, in the middle of June, I have found the temperature of Cintra nearly 20° lower than that of Lisbon.

The climate of Lisbon is warm, but humid and variable. The mean annual temperature is about 61°; the mean temperature of winter is 54°; spring, 59°; summer, 68°; and of autumn,

59°. The minimum temperature of the year is seldom under 35°, but occasionally a slight frost occurs during the night in December. The maximum annual heat in the sun is towards the end of August, when the mercury at 2 P.M. sometimes rises to 133°. The extremes of heat and cold annually observed differ by between 50° and 60°.

These thermometrical details are deduced from the scattered observations of various native and foreign observers, from some tables quoted by Mr Murphy, and especially from two papers by Senhor Franzini, in the Transactions of the Royal Academy of Lisbon,* all of which I have tabulated together.

No sufficient register of the hygrometric state of the atmosphere has been kept, but it is well known by every observer that the air in winter is here laden with vesicular moisture. Senhor Franzini states that humid winds predominate over dry ones in the proportion of 199 to 165.

Rain usually falls in heavy showers of brief duration. Thus the number of days of settled rain in the year are comparatively few, not exceeding 63, while the amount of rain is considerable, amounting to upwards of 30 inches annually. One-half of this quantity falls in winter, from November to the middle of February; the remainder being nearly equally divided between spring and autumn, as there is little or no rain from June to October. During the rainy weather the south-west wind is prevalent.

To the stranger recently landed in Lisbon from England, in winter, the weather will at first appear by comparison mild and genial; and so indeed it very often is. But still at this season cold and damp winds from the westerly points are prevalent. And although frost is a rare phenomena in the

* "Observaçoes Meteorologicas Feitas na Cidade de Lisboa, &c., par Marino Miguel Franzini, in Memorias da Acadamia Real Das Sciencias de Lisboa, tom. v. p. 92, 125, et tom. vii. p. 61, ad 93.

town, yet the nights in the large fireless rooms of Lisbon are often chilly and uncomfortable, and I have more than once seen entire families muffled up to the eyebrows in every procurable wrapper, cloak, and even blanket, shivering over the *brazeiro*. But such weather is an exception to the usual mildness of the winters, the temperature of which, though much below that of several other health resorts, is, however, ordinarily fully as many degrees higher than in the most favoured of our home winter climates. In spring, however, cold winds, accompanied with a very hot sun, are prevalent.

Of late years Lisbon has been little resorted to as a winter residence by British invalids, although formerly it was one of the most frequented of these places.

Among the diseases to which the climate of Lisbon appears applicable, may be included chronic winter cough with increased sensibility, and permanent sub-inflammatory condition of the pulmonary mucous membrane. This state, however, will demand a very careful diagnosis, bearing in mind those forms of phthisis in which the climate would prove injurious. Dyspepsia, attended by a similar state of the mucous membrane of the stomach, will sometimes be cured by a change from England to Libson.

That the climate of Lisbon acts unfavourably on the physical development and constitution of young persons, is, I think, established by the stunted and prematurely aged appearance of the Portuguese children after the age of fourteen. Up to this period they thrive equally well, and even grow faster than do children in this country, but once passed that age their growth becomes arrested, and their carriage and aspect are those of old men, before they attain puberty.

In some measure, however, this must result from the excess to which smoking is carried, even by young children. I have often seen, with astonishment, boys whose age could not have exceeded six or seven years, gravely sucking a strong cigar, with apparently the same gusto which our 'less precocious infants derive from the forbidden delights of the sugarstick. There can be no doubt that the influence of the nicotin thus absorbed must be most injurious at this tender age. But even irrespective of this, it would seem that the climate is unpropitious to youth.

CHAPTER VII.

MADEIRA.

A FEW years ago Madeira was *par excellence* the winter resort
of English consumptive patients. Now, however, this island
is comparatively deserted by such persons, and having been
formerly recommended in cases in which it was not suitable,
has now come to be almost disused even where it might be
serviceable. Moreover, the tide of fashion, which influences the
choice of health resorts so largely, has for the present set in
other directions. The remoteness and comparative inacces-
sibility of this island, thirteen hundred miles from the nearest
English port, has also conduced to the preference now generally
given to nearer winter resorts.

From Southampton or Liverpool, Madeira may be reached in
about eight days by the steamers of the African, Union, or
Brazilian Mail Companies.

A remarkable illustration of the vague manner in which
health resorts are sometimes recommended may be found in
the last edition of Dr Constantine James' work on this subject,
in which he speaks of Madeira as the "Type des climats
d'hiver," * without distinguishing the fact that it is the type
only of one class of climates, by no means generally applicable
for pulmonary invalids.

---

* "Guide Practique des Eaux Minerales et aux Stations Hivernales," par
le Dr Constantine James, IXme edition, p. 573, Paris, 1875.

Of all the climates described in this work, that of Madeira is unquestionably the most equable. Thus, during winter and spring, the mean range of temperature observed at Funchal was only 15° in the twenty-four hours, whilst in Upper Egypt there is an average range of 40° during these seasons. But still the latter place is incomparably better suited for the majority of phthisical invalids than the former, the atmosphere of Upper Egypt being very dry and tonic, whilst that of Madeira is essentially humid, and in many instances relaxing.

The extreme humidity of this climate is shown by the impossibility of keeping steel instruments free from rust, or of preserving any musical instrument in tune, or any article of clothing, however carefully packed, from being injured by the dampness of the air; as well as by the exuberant tropical vegetation, which attracts the admiration of every visitor to this island, and which, as rain falls only in small quantities, and at very long intervals, must be maintained by the excessive humidity of the atmosphere.

The mean annual temperature of Funchal is 66°·93, and according to Dr Mason's observations,* the minimum external temperature was 57° in January, and the maximum was 83° during the prevalence of the hot African "Leste" wind in June. Independently of this wind the maximum day temperature was 79° in August.

External Temperature, Funchal, Madeira.

	Mean Maximum Temperature in the day.	Mean Minimum Temperature in the night.	Mean Daily Range.	Mean Range of the Twenty-four hours.
Winter, . .	68·66	55·00	7·60	13·66
Spring, . .	74·59	57·50	9·50	17·00
Summer, . .	80·00	65·00	8·67	15·00
Autumn, . .	76·33	61·66	8·33	14·67
Year, . .	74·87	59·79	8·54	15·08

* "A Treatise on the Climate and Meteorology of Madeira," by the late J. A. Mason, M.D., p. 183.

D

The prevailing wind in Madeira is the north-east, except at Funchal, where, owing to the position of the adjoining mountains, southerly and south-west winds are more frequent. Throughout the island generally, land and sea breezes alternate regularly, and mark the changes of the seasons by the order in which they occur, the one prevailing during the night, and the other from sunrise until evening. In summer the "Leste" wind from the African coast occurs periodically for a few days, and is very similar to the "Terral" of Malaga, or the "Khamsin" wind of Cairo, being not only intensely hot and dry, but also laden with impalpable sand, and most irritating to the lungs and injurious in its general effects.

It has been proved that "the inhabitants of Madeira are not remarkable for longevity, but, on the contrary, in general die very young;" and though this fact has been explained in various ways, it is nevertheless a strong *a priori* argument against this island as a health resort. Amongst the prevailing diseases of Madeira, consumption must be included; but whether the prevalence of this disease amongst the native population is to be ascribed to the climate, or to the unfavourable hygienic condition and meagre diet of the poorer classes, has long been a disputed question. The results of the experiment made a few years ago by the authorities of the Brompton Hospital were by no means favourable to the reputation of Madeira as a health resort for the consumptive. Twenty-six carefully-selected cases of phthisis were sent to winter in Madeira; of these only two were decidedly improved, seven were slightly improved, one died from hæmoptysis, five returned worse than when they left home, and in twelve cases no alteration could be observed in the patients' condition.

The cases in which a winter visit to Madeira might prove serviceable are—in dry chronic bronchitis and winter cough, with great irritability of the mucous membrane of the air passages and little expectoration ; chronic laryngeal irritation, and cases of spasmodic asthma of the same character ; and also, with more caution, in a limited number of cases of phthisis in its earliest stage, and marked by similar symptoms. But to the majority of consumptive patients, especially if the disease be advanced beyond the first stage, the humid, warm, and relaxing atmosphere of Madeira would, I believe, be prejudicial.

The foregoing account of the climate of Madeira might, with very little alteration, be applied to two other groups of not far distant islands, also situated in the North Eastern Atlantic Ocean, and which have been as deservedly recommended as health resorts in the same class of cases as Madeira, viz., the Canaries and the Azores. Some years ago I had an opportunity of visiting both these groups. The best known of the Canaries—Teneriffe—lies some two hundred and fifty miles to the south of Madeira, and is only sixty miles distant from the African coast. This island, the circumference of which is less than a hundred and fifty miles, has been so often described by tourists, and is now so little visited by health-travellers, as to render any detailed account superfluous. The voyage by steam from Liverpool to Teneriffe varies from eight to ten days, and costs about twenty pounds. There is pretty good hotel accommodation at Santa Crux, the capital of the Island. The mean annual temperature of this town is 6° higher than that of Funchal. But the climate is much less equable than that of Madeira ; thus, according to Sir James Clark, the difference between the mean temperature of the winter and summer at Funchal is nine degrees, whilst at Santa Crux it amounts to twelve degrees. The mean annual temperature is 71° ; that of winter 65° ; spring, 69° ; summer, 77° ; and autumn, 74°. The annual rainfall is somewhat less than that of Madeira, nor is the atmosphere quite so humid and relaxing. But its proximity to the African coast exposes this island to harsh dry easterly winds from the Lybian deserts; and, moreover, the distance from England, and comparative deficiency of hotel accommodation and resources for foreign visitors, render this island still less adapted as a health resort for British pulmonary invalids.

The Azores being two hundred and fifty miles to the northwest of Madeira, enjoy a cooler but more humid, relaxing, and equable climate than that sanatorium. The principal of these islands as a health resort is St Michael's, the mean annual temperature of which is five degrees lower than that of Funchal; and that of winter two degrees under that of the latter town. The prevailing winds in St Michael's are northerly and easterly; the annual rainfall is about thirty inches, and the atmosphere is still more saturated with moisture than that of Madeira. I need not dwell further on the climate, which, although one of the warmest and most equable in the world, is far too relaxing for the great majority of pulmonary sufferers, though it might advantageously be employed in some cases of spasmodic asthma, dry chronic bronchitis, and laryngeal irritation, were it not for the disadvantages of a health resort so remote, difficult of access, and unprovided with many of those comforts and resources which are regarded as indispensable by invalids.

CHAPTER VIII.

ALGIERS.

FEW of the ordinary winter resorts of invalids are more easily approached from England than Algiers, as there are regular departures from Marseilles every second day, for that city, by the steamers of the "Messageries Maritimes" and those of the "Compagnie Valery." This voyage, in fair weather, and under ordinary circumstances, is a mere pleasure excursion, occupying from thirty-six to forty hours.

The first view of Algiers from the bay is very picturesque ; from the water's edge a triangle of houses extend to the top of the hill, the apex being crowned by the ancient fortress of the "Casbah," the last refuge of the Deys. The snow-white walls of the city glistening under an African sky, the lofty minarets still surmounted by the crescent, the grave turbaned figures we see loitering about the mole,—all give an oriental character to the scene.

There is a wide choice of hotels, of which the best seemed to us, after having tried others, the Hotel d'Europe, where we remained the greatest part of our stay in Algiers. The usual charge in all of them is from twelve to fifteen francs a-day, including breakfast, dinner, and apartments.

The markets are well supplied, and the prices are somewhat less than in this country, but the meat is not equal to ours. Fruit is abundant; and as a proof of the mildness of the climate, I may mention that during December we had green peas and fresh strawberries, grown in the open air, almost every day at dinner.

The streets in the higher quarters of the city still retain their oriental aspect to a remarkable extent, being tortuous, dirty, and so narrow that in many two horsemen can with great difficulty pass each other. When a string of donkeys laden with projecting sacks appears in one of these lanes, all the passers-by are forced to retreat into the nearest door-way, and woe betide the luckless individual who fails to secure this timely shelter.

In the modern part of the city, near the port, the streets are wide, well paved, and kept perfectly clean. Nearly all these communicate with the Place du Gouvernement, which over-looks the harbour, and around which most of the hotels and cafés are situated. The chief thoroughfares, such as the Bab-el-Oued and its continuation, the Rue Bab-Azoun, the Rue de Chartres, &c., are well lighted with gas, and contain shops that would not discredit Paris itself. All these streets are built with arcades overhanging the footpaths, and thus a dry walk, in wet weather, and a cool retreat from the scorching sun in summer, is secured. By invalids, however, these colonnades, in which even on the hottest day there is always a cold draught, should be carefully avoided, and are most injurious.

In the way of amusements there is little to tempt the stranger, and even the gay and mercurial French seem to be easily affected by the hot Algerian climate. They soon become bilious and gloomy, and appear to consider only how they may most quickly amass sufficient means to enable them to leave the colony and return to France. A large and handsome theatre has, however, been erected in the Place Bresson, but is very poorly supported.

One of the chief advantages of Algiers consists in the

number of opportunities for agreeable excursions from the town, and the beauty and variety of these.

To antiquaries Algiers presents an almost virgin field of research and discovery; rich in monuments of Carthaginian and Roman antiquity, and in remains, too, of a date long antecedent to either. A paper on some of these, read before the Royal Irish Academy, by my father, Dr R. R. Madden, contains, I believe, the first account given of "The Dolmens of Bainam."

According to M. de Pietra Santa, the mean annual temperature of Algiers is 66°. The mean temperature of winter is 55°; spring, 66°; summer, 77°; and autumn, 62°. During a period of seven years, Dr Armand * tells us that the thermometer only once fell to zero.

In considering the climatology of Algiers, I have adapted Dr Armand's division of the year into two seasons, the first that of the rains, commencing in November and terminating at the end of April, and the second that of the heats, which endures from May to October, inclusively.

Temperature of the Coast of Algeria during the cool and wet season.

	Maximum.	Minimum.	Mean.
November,	68°	57°	63°
December,	60	50	55
January,	60	50	55
February,	54	46	50
March,	64	55	58
April,	70	58	64

During these months snow falls in considerable quantities on the mountains, at the height of 500 metres above the level of the sea. Snow very rarely falls in the valleys along the

* "L'Algerie Medical," p. 49.

coast, and when it does fall, generally thaws as soon as it touches the ground.

The effects of the extreme cold of the African mountains were dreadfully experienced by the French troops at the retreat of Constantine, which, according to Dr Armand, a physician attached to the French army, "resembled on a small scale the horrors of the retreat from Russia. The column of Setif, on the 2d and 3d of January, were assailed and dispersed by a snow-storm in the mountains of Bou-Taleb, and entered Setif on the 4th, with 532 cases of partial congelations of the extremities, and leaving 208 dead on the road." *

During the month of December, the thermometer in the open air and in the shade, in a westerly aspect, was very carefully observed by me in Algiers at regular times, twice daily, and also at various hours between 4 A.M. and 11 P.M. At no period of the day did the heat ever exceed 68½°. The mean temperature at noon during the month was 66°. The mean temperature of the night was 58¼°, and the lowest temperature observed at any period of the day or night was on Christmas day at 4 A.M., when the temperature was 56°, and at noon was only 60°. The mean variation up to the 24th was 7¼°, and the greatest variation during the twenty-four hours was on the 14th December, when the temperature at noon was 68½°, and at 8 P.M. had fallen to 59°, being a variation of 9½°.

In 1738 Dr Shaw, who for twelve years, while chaplain to the English Factory, kept a very accurate meteorological register at Algiers, tells us that the average annual rain-fall was 27 or 28 inches.† It would seem from this that the climate had become damper since his time, for, according to nearly all the modern observers, the annual rain-fall now

---

* "L'Algerie Medical," p. 42.

† Shaw's "Travels," Oxford, 1738, folio, p. 218.

amounts to close on 32 inches, and the average number of rainy days in the year is from 55 to 60.

Six-sevenths of the annual rain-fall, and seven-eighths of the total annual number of rainy days, occur between November and April inclusively. These rains occasionally give rise to veritable inundations. Fortunately, however, these are seldom very serious, and the floods oftentimes prove beneficial in their after effects, fertilising lands that would otherwise be parched up into a barren desert.

In summer, notwithstanding the sea-breeze which generally prevails along the coast, the thermometer attains a height unknown in any part of Europe, accompanied by a sultry stillness of the air, destructive to vegetation and injurious to animal existence. Such was the weather which the poet Campbell, in his epistle to Horace Smith, alludes to—

"Dear Horace! be melted to tears,
For I'm melting with heat as I rhyme ;
Though the name of this place is All-jeers
'Tis no joke to fall in with its clime."

The hygrometric state of the air differs materially in the lowlands and mountain districts, being generally dry in elevated places ; while in the valleys the atmosphere during the night is saturated with moisture, giving rise to damp and foggy weather, which, occurring after the intense heat of the day, occasions much mischief, and partially explains how it is that fever is endemic in some localities.

Westerly winds are most prevalent in Algiers. Those of the east are next in frequency, and are generally attended with broken weather and cold mists.

At all seasons the temperature of this place is subject to great and sudden changes. Profuse dew falls at night, and the atmosphere after sunset becomes loaded with moisture. For this reason it is never safe for valetudinarians to be abroad before the sun has some power in the morning, nor after sunset at all. Exposure to the night air is not only dangerous to European invalids, but prejudicial to persons in health. Throughout the whole of Algeria there is a great tendency to febrile affections, though they are less common in the city of Algiers and the adjacent villages than in any other part of the colony. Wherever there are marshes, or lakes, or when soil, not previously cultivated, has been cleared for the first time, there intermittent fevers prevail. In chronic cases a remarkable enlargement of the spleen takes place, that organ, as I have seen in several instances in the civil hospital of Algiers, apparently occupying nearly the entire abdominal cavity. These *fièvres paludèens*, have been the bane of Algeria, and owing to this cause the population has in some localities been renewed three times since the French colonization. This tendency to malarial fever renders great caution necessary on the part of invalid visitors. I myself suffered from a severe attack of this kind, following fatigue and exposure to the sun during a short excursion on the verge of the desert. This assumed an intermittent type, and, even long after my departure from the colony, recurred again and again on the slightest exciting cause.

The Algerian climate is very fatal to European children between the age of six months and two years; the mortality at this period of life, according to Drs Foley and Martin, authors of an official work intended to encourage emigration to the colony, amounting to 440 per 1000, or nearly one-half. "This," they justly observe, "is enormous, and indicates the almost impossibility of rearing in Africa European children, who have been brought over during dentition."*

* "Histoire Statistique de la Colonisation, Algérienne au Point de vue du Peuplement et de L'Hygiène," p. 107.

From the same valuable work we also learn how greatly the death-rate of this colony exceeds that of the mother country.

Affections of the eyes, especially scrofulous ophthalmia, are common, and so is scrofula in all its forms. In the civil hospital at Mustapha, the physician pointed out to me several well-marked cases of scrofulous disease of the bones. He accounted for the prevalence of this malady by the very poor food on which the lower class of the natives subsist, and partly, also, by the crowded state and defective ventilation of their dwellings.

It has long been a disputed question whether pulmonary consumption is prevalent in Algeria or not; and also whether this climate is a suitable one as a health resort for the phthisical, or the contrary. Nearly forty years ago, when this matter was brought under the consideration of the French Academy of Medicine, it was declared that—"it was doubtful if the climate of Algiers was favourable to the cure of consumption;"* and even at the present time little more can be said.

The prevalence of scrofula supports the statement of Drs Armand,† Deleau, and Laveran,‡ physicians of the French army in Algeria, that phthisis is common among the native Arabs. This opinion is also maintained by the Tebibs, or native doctors, who, moreover, believe phthisis to be contagious.

Many writers, however, consider that consumption is all but unknown in Algeria, and that, when it does occur, it is only the development of disease existing before arrival in the colony, or inherited from phthisical parents. Dr Odrultz,§

* "Bulletin de L'Academie Royal de Medicine," 1836, vol. i. p. 43, &c.
† "L'Algerie Medical," p. 375.
‡ "Memoires de Med. Milit." t. lii.
§ "Annuaire Therapeutique de M. Bouchardat."

Dr Bodichon,* Dr De Pietra Santa,† Dr Feuillet,‡ and some other recent writers support this opinion.

From what I saw myself during my stay in Algiers, I have no doubt that consumption is by no means unfrequent amongst the resident population, although it is less common than in England or France, and that other diseases of the respiratory organs are prevalent in this colony. Thus, out of five hundred and forty patients in the civil hospital of Algiers when I visited it, no less than fifty were pulmonary cases. And I believe that unless great precaution be observed by invalid visitors to Algiers, in their dress, mode of living, and above all in the careful avoidance of exposure to the night air and cold winds, and a constant recollection of the remarkable difference of temperature which exists between the sun and shade, this climate is a hazardous one for persons suffering from pulmonary complaints.

Cases of phthisis, accompanied with hectic sweats, are often aggravated here, and I have heard invalids complaining that their night perspirations, which had been controlled by change of air in the journey from England, again returned after a few days' residence in Algiers.

For a patient in the third stage of consumption it would be difficult to select a more unsuitable climate than this, or one where the disorganization of the pulmonary structure and death of the patient would be more likely to be accelerated.

Some cases of chronic bronchitis derive remarkable benefit from the climate of Algiers. Several instances fell under my observation of elderly persons suffering from this disease, and from senile catarrh, who here regained, to an unexpected extent, their general health, after having gone the rounds of

* " Algeria as a Winter Residence for the English," p. 44.
† " Journal of Practical Medicine and Surgery," vol. xxxi. No. 10, p. 469.
‡ " De La Phthisie Pulmonaire en Algérie," 2d edition, p. 15.

all the wintering places in the south of France and Italy without any amelioration. But other cases of the same disease have also come under my notice, in which, the climate of Algiers having proved ineffectual or injurious, Malaga was successfully resorted to.

CHAPTER IX.

MOROCCO.

WITHIN four hours' sail of Gibraltar, with which there is
almost daily communication by steamer, is an empire almost
as large as Spain, no small part of which is blessed with a
most genial climate, and possesses the most fertile soil in the
world. Extending from the Atlantic on the west, to Algeria
on the east, and from the Mediterranean stretching far into
the Great Sahara on the south, Morocco includes a territory
of upwards of 150,000 square miles, a considerable part of
which is to this day almost a *terra incognita.*
    The following brief account of my visit to Morocco may
perhaps serve to direct the attention of others to this climate,
and its possible therapeutic influences :—Arriving at Tangiers
by steamer from Gibraltar, the town was not seen until we
were close to the land, as, although built on a hill, the houses
are for the most part flat-roofed and low.   This hill is
surmounted by an old Moorish Casbah or castle, whose
dilapidated white-washed walls would make a poor defence
against modern artillery.   Below this rises the lofty minaret
of the Jamaa Kiber, or principal mosque, while nearer the
port the forest of masts made it appear in the distance as if
some crowded dock was placed there.   These, however, were
but the flag-staffs of the various consuls, who each deem it a
point of national honour to have a taller pole than their

neighbours. All this presented such a contrast to the place we had just left, that it was difficult to imagine that Gibraltar was only thirty-eight miles distant. The steamer was forced to anchor a long distance from the land, as the port was destroyed when the English evacuated Tangiers in 1684, and has never been restored since.

The town is surrounded by semi-ruinous walls, passing through which we entered the main street; this traverses the city from east to west, and is tolerably wide and clean, containing the principal shops and bazaars, as well as the Jamaa Kebir, or great mosque, a large, but very ugly, brick building, a little above which the street expands into the market-place.

There are some four or five hotels in Tangiers. These for the most part are comfortable and very moderate; thus in the best of these we lived very well for rather less than seven shillings a day.

While we were in Tangiers the annual caravan of Hadjis, or pilgrims to Mecca, arrived from the interior, and remained here for some days, waiting for the steamers by which they were to perform their voyage to the East. Amongst these pilgrims were men of every hue, from the perfectly white and dignified Moor of El-Garb, clad in a rich and most becoming costume, to the jet black native of Soudan, whose only clothing was a narrow cloth around the loins, and perhaps a ragged blanket thrown loosely over one shoulder.

In the encampment of these Hadjis, outside the walls of Tangiers, we witnessed a most extraordinary performance. In the centre of a great crowd of Hadjis were a party of four or five negroes, dancing and screaming to the music of a kind of rude tambourine. Suddenly the music, if the noise could be so called, stopped for a minute, and one of the performers came forward holding a stick, to the end of which a large clasp-knife was fastened in one hand, and having in the other a cannon ball, weighing about ten or twelve pounds ; which was handed round and examined by several of our party. The dancing and shouting recommenced, and the actors gradually worked themselves into a state of extreme excitement; the chief now sat down in the centre of the ring, and poising the ball in his right hand, threw it by a powerful effort several feet into the air, and stooping forward received it in a slanting direction on the upper part of the frontal bone, the force of the blow producing a dull sound which was distinctly heard by every one in the circle; and the blood, now oozing in streams from the bare scalp, covered the man's whole face. After rushing round and round, singing and dancing for some moments, he repeated the same thing three or four times, and then fell senseless to the ground; the others dancing about him until he came to, when he at once commenced cutting himself about the head and face, with the knife I have before mentioned, and then, when his features were so gashed and disfigured as to be hardly discernible, he threw himself down and remained lying on his face, biting the earth until I came away, one of the performers meanwhile going about soliciting money, of which he got very little. Close by was a rival exhibition of snake-charmers, who were freely handling large serpents.

How the first of these feats was accomplished I am utterly unable to explain, for had the same blow been given to any European cranium, I have no doubt that a depressed fracture would have been the result; and even supposing that this negro's skull was of sufficient thickness to sustain the force of such a blow without serious mischief, I cannot conceive how the repeated shock of so great a concussion could have been borne without occasioning fatal injury to the brain.

The environs of Tangiers are interesting, especially the cliffs

on the west of the town, which command a very extensive
and beautiful view of the straits and Spanish coast from
Trafalgar to Gibraltar. Here the outlines of old English
batteries may still be traced; and half-buried in the sand we
found several rust-consumed cannon, stamped with the
English arms of the time of Charles II. On the south side of
the city ruins of a very remote antiquity are discernible;
these are supposed to have belonged to the ancient Tingis;
and in the same neighbourhood several of the arches of an
extremely graceful Roman bridge stand in tolerable preservation. Outside the land gate is the *Sok-el-wahad*, where the
principal market is held every Sunday; beyond is the
neglected and ill-kept cemetery, above which, on the hill, is a
white bee-hive shaped *coubba* or Santon's tomb.

The natural history of northern Morocco differs little from
that of Andalusia, the only difference perceptible being that
the soil in Barbary is far more productive than it is on the
other side of the straits, although the cultivation is of the
rudest kind. In the province of El-Garb two crops in the
year are of ordinary occurrence, and Gibraltar depends entirely on Morocco for its excellent supply of oranges, lemons,
grapes, figs, melons, and dates, which are all sold here at
fabulously cheap prices.

No very accurate details of the climate have, I believe, been
yet published. During my stay in spring the weather was
mild and genial, and at mid-day was even too hot for much
exercise. The position of the town, however, exposes it completely to the cold damp winds, which rush through the
funnel-shaped straits from the Atlantic, while its aspect,
being open to the east, must render it even more subject to
this wind than Gibraltar is. The annual rainfall is about
thirty inches, which, as in most parts of Africa, principally
occurs at one season, during the months of October and

E

November. The rains being succeeded by great heat, vegetation is consequently rapid and early; thus in January the fields are already covered with flowers, and in March the barley crop is reaped. Though the climate is hot it is not parched or arid, as the province of El-Garb is protected by the interposition of the two ranges of the greater and lesser Atlas mountains on the south and south-east from the hot winds of the desert. Its proximity to the Mediterranean and Atlantic on the north and west also modifies the temperature, which in this province seldom falls below 40° in winter, or rises above 86° in summer.

From the climate we next come to examine the diseases to which it gives rise. Ophthalmia is very prevalent in Tangiers; so is catarrh, especially in the Jewish quarter, where many of the inhabitants present a pallid, unhealthy, and even phthisical appearance. Elephantiasis Arabum is common, and I have seen a mendicant outside the gate of Tangiers exhibiting a leg enlarged by this disease to nearly three times the size of its fellow.

Mr Jackson, who resided for sixteen years in the Barbary States, asserts that leprosy, called _Murd Jeddem,_ is endemic throughout Morocco;* and Mr Lempriere says that "the leprous affection appears to be hereditary, and has the appearance of being the true leprosy of the ancients. It breaks out in great blotches over the whole body, in some few forming one continual sore, which frequently heals up, and at stated times breaks out afresh, but is never thoroughly cured." †
Constitutional syphilis, or _Murd Kibeer,_ is remarkably widespread throughout this country, and the true Oriental plague has repeatedly almost depopulated Morocco.

* "An Account of the Empire of Morocco," &c., p. 154.
† "A Tour from Gibraltar to Morocco," by William Lempriere, Surgeon, p. 9, 2d edition.

To this long catalogue of ills must be added every form of scrofulous disease, and during the stay of the Hadj in Tangiers I had an opportunity of observing the number of persons evidently of the strumous diathesis who were here collected together from every part of the empire. This must, to some extent, be attributable to the utter neglect of hygiene and poverty of diet of the lower classes. But still, the mere prevalence of scrofula in any country, from whatever cause it may arise, should make us most cautious in selecting that place as a health resort for the consumptive.

I have not, however, had sufficient personal experience of this climate, in affections of the respiratory organs, to enable me to speak absolutely of its action in these cases. Several invalids who were of our party seemed to derive benefit from their stay in Tangiers; this, however, was probably owing to the superiority of the living here to that of Malaga, where they had been spending the winter. For the same reason I think that invalids suffering from dyspepsia and hypochondriasis, and possibly persons in whom climacteric disease has begun to manifest itself, as well as that large class of whom I have treated in the beginning of this work, namely, valetudinarians, not actually ill, but whose state is best described by the vague term, " out of health," may in some cases derive advantage from a short visit to Tangiers.

At right angles to, and only fourteen miles distant from, Gibraltar, stands its formidable rival, the Spanish key of the Mediterranean—Ceuta, which, like Gibraltar, is built on a rocky peninsula jutting out into the straits, of which it forms the south-eastern point. This peninsula, nearly three miles in length, terminates in the bold headland of El Minah, and is formed by seven hills. One of these mountains, now known as Mount Hacho, is supposed to be the Abyla, or southern pillar of Hercules.

In most respects Ceuta resembles its opposite neighbour, its situation beingvery similar, being, like it, a small, strongly-fortified town and convict settlement in a foreign land, and, notwithstanding its extreme antiquity, is almost as destitute of any object of interest within its walls. The environs, however, are strikingly beautiful and wonderfully fertile, producing excellent oranges, as well as figs, melons, grapes, and sugar-cane.

The climate of Ceuta differs little from that of Gibraltar, although somewhat warmer and better sheltered from cold, harsh winds. In this respect it is also more advantageously situated than Tangiers.

Tetuan or Tetawan is about twenty-five miles south of Ceuta, and is nearly four miles inland. It was formerly a place of some importance, but has now fallen into decay and ruin. The environs are highly cultivated, however, and produce the finest fruit in the world.

The climate is similar to that of Tangiers, though in one respect superior to it, being in a great measure sheltered from the damp winds to which that place is exposed. It would, however, be impracticable for any invalid to remain in this town, from the total want of all the conveniences of life, as well as from the difficulty, not to say danger, of living amongst an uncivilised and fanatical Mahometan population, who regard all Europeans with contempt and aversion. The same observation applies equally to the much superior climate of El Araiche, a very ancient Moorish town, situated just within the mouth of the river Kos, about forty-five miles south of Cape Spartal, which may be reached by a journey of a day and a half from Tangiers, and which has also been recommended as a winter residence for invalids ; but on account of the greater difficulty of access, and total want of accommodation, is still more unsuitable as a health resort for European valetudinarians.

CHAPTER X.

PAU, ARCACHON, AND BIARRITZ.

WITH one exception, the most frequented winter health resort in Europe is Pau. In the following chapter, founded on my own experience during two visits at long intervals, I shall endeavour to enable the reader to judge whether the reputation which each year attracts so large a number of British invalids to this place is well founded or not.

Pau, the ancient capital of Bearn and Navarre, which may be reached in eighteen hours by train from Paris, *via* Bordeaux and Dax, is situated on a small table-land about 700 feet above the sea, in the department of the lower Pyrenees, one hundred and fifty miles from Bordeaux, and twenty miles from the nearest part of the Pyrenees. The platform on which the town stands is intersected by a deep ravine, through which a scanty and very dirty rivulet flows, "stealing and giving odours." On the north of this ravine is the new town, in which are many of the public buildings, and most of the houses inhabited by foreign visitors.

The old town, on the opposite side of the river, as seen from the bridge, reminds one somewhat of Edinburgh on a small scale. In this quarter are the Prefecture, the Chateau Henri IV., and the Place Royal, where stands the beautiful statue of Henry of Navarre, erected by Louis Philippe. This promenade commands an exquisite view of the rich valley of the Gave, and an unbroken vista of the Pyrenees for fully sixty miles. The streets of Pau are dirty and badly paved, but the shops are well supplied with goods suited for English customers. The hotels, especially the Hotel Beau Séjour and the Hotel de la Poste, are comfortable and not expensive. Apartments may also be had at all prices, from one thousand to ten thousand francs for the season from October to May. It is essential to choose rooms commanding a northern aspect, and having fire-places, as there is a difference here generally of 10° or 12° between the north and south side of the same house, and the nights are always, and the days generally, cold enough to render a fire necessary in winter. In most of the houses the sewerage is very defective, and the effluvia in the halls and staircases, especially in wet weather, may be better imagined than described.

The climate of Pau has been written up very assiduously, and highly extolled for its peculiar advantages during eight months of the year, including the entire of the winter, as an especially well-adapted place for invalids labouring under diseases of the respiratory organs. But no amount of eulogy affects the temperature of a locality, or improves the hygro-metric condition of its atmosphere. And, therefore, though the books that have been written on Pau have answered their purpose most successfully, by annually bringing crowds of invalid travellers to spend the winter months in a locality in favour of which they have read such strong and circum-stantial statements, still the climate remains what it was be-fore the books were written—essentially cold, variable, damp, and dreary during the winter.

I should feel some hesitation in offering so unfavourable an opinion of this climate, which is almost universally regarded as one of the finest in the south of Europe, if I had not had some personal experience of its action on invalids in whom I was interested. Moreover, I have been favoured with very extensive unpublished meteorological and other observations, a résumé of which will be found in this chapter.

Confiding in the works on this climate I had read, I arrived in Pau from Algiers in the beginning of January, expecting to find a mild, equable, and genial climate. I was soon undeceived, however; the change was literally from an atmosphere warm and bright as summer in this country, to one nearly as cold and damp as that of London in mid-winter. The evenings were foggy and the mornings misty, and during the month of January there were few days on which any person suffering from pulmonary disease could go out of doors without prejudice to health. At Algiers, during the month of December, the thermometer as observed by me never fell below 54° at any hour of the day or night; while at Pau during the same month, according to the observation of M. Weil, the mean temperature of the day was considerably under the lowest temperature of the twenty-four hours at Algiers, being 45½°, and the thermometer fell eleven times to zero.

Probably the best mode of studying a climate is by comparing it with others, and therefore I will now contrast Pau with Dublin, which may be considered a specimen of a bad climate; and likewise with Malaga, which I regard as a very good one. The mean annual temperature of Pau is 56°, or 7° higher than that of Dublin and 9° lower than that of Malaga. The mean temperature of winter is 42°, or 3° higher than Dublin and 13° lower than Malaga. The mean temperature of spring is 54°, or 7° higher than Dublin and 14° lower than Malaga. In summer the mean temperature of Pau is 70°, or 11° higher than Dublin and 8° lower than Malaga. And, lastly, the mean temperature of autumn is 58°, which is 8° higher than that of Dublin and 2° lower than that of Malaga.

From this we learn that in winter, when this town is crowded by English invalids, the temperature of Pau is only 3° higher than that of one of the worst climates in Great Britain, and is 13° lower than that of Malaga. In spring, however, the climate is somewhat better.

The following tables, the first of which is copied from the registry in the club at Pau, and the second arranged from a valuable register kindly given me by M. Weil of that town, will afford the reader some idea of the character of this climate.

Table showing the comparative Meteorology of Pau and of Kew Observatory during an average Winter.

	October.		November.		December.	
	Pau.	Kew.	Pau.	Kew.	Pau.	Kew.
Mean temperature at 9 A.M.,	60·5	55·6	44·7	41	41·7	40
Mean of the highest temperature by day in each twenty-four hours	70·5	61·6	57·3	46·6	52	45
Mean of the lowest temperature by night in each twenty-four hours	52	48	41·8	33·2	37·9	34·8
Mean temperature of the months calculated from preceding observations,	60·4	55·1	49·1	39·5	45	40
Number of days in which the thermometer fell to freezing point,	0	0	1	13	5	11
Number of days on which rain fell	7	10	14	20	8	11
Amount of rain in inches and hundredths,	1·28	0·74	2·734	1·054	1·275	0·925
Mean amount of cloud,	4·2	7	5·3	6·1	4	6
Mean moisture of air,	·79	·85	·82	·85	·83	·83

Pau is one of the most variable climates in the south of Europe. Thus, for instance, I have seen a difference of 20° between nine A.M. and noon. I need hardly add that such sudden changes of temperature must be injurious to consumptive patients.

It must be admitted, that if we contrast the comparative number of rainy days at Pau and Dublin, the advantage will be on the side of the former. Thus, in Dublin, the average annual number of wet days is 224, while in Pau there are only 119; yet in Pau " it never rains but it pours," for the actual quantity of rain which falls is 42 inches, or 15 inches more than at Dublin, and 26 inches more than at Malaga, where there are only about 30 rainy days annually.

Temperature at Pau.

Date.	December.			January.			February.			March.		
	9 a.m.	Noon.	3 p.m.	9 a.m.	Noon.	3 p.m.	9 a.m.	Noon.	3 p.m.	9 a.m.	Noon	3 p.m.
1st	43	52	54	...	...	...	...	...	...	44	53	55
2d	41	52	54	41	48	48	44	46	50	42	53	57
3d	37	52	52	41	46	44	46	50	50	44	57	59
4th	37	50	52	33	41	39	41	44	42	39	41	44
5th	36	45	46	41	44	44	39	41	39	41	46	50
6th	43	50	50	41	46	44	39	41	44	42	53	59
7th	54	55	55	41	46	46	37	41	37	44	59	64
8th	43	54	55	39	44	46	32	35	35	50	66	71
9th	45	52	54	46	48	53	24	24	26	53	69	71
10th	49	54	55	50	59	59	21	28	30	48	53	46
11th	49	54	54	42	53	55	21	28	35	44	48	48
12th	41	54	55	41	50	50	23	35	35	44	46	46
13th	43	52	54	46	44	42	24	37	37	42	41	41
14th	43	49	50	41	41	41	28	48	48	41	50	51
15th	41	52	54	39	41	41	48	50	50			
16th	43	52	49	37	39	37	37	48	48			
17th	43	46	45	26	35	37	37	48	48			
18th	43	45	45	26	30	33	39	50	48			
19th	41	45	43	26	35	37	37	41	44			
20th	37	37	40	32	39	41	41	44	48			
21st	40	41	41	32	48	48	39	46	42			
22d	34	43	46	41	46	48	41	44	46			
23d	37	49	49	42	50	51	37	39	42			
24th	36	46	49	48	57	57	42	50	50			
25th	36	46	46	46	53	50	37	41	39			
26th	36	50	49	46	51	50	42	48	50			
27th	36	43	43	35	50	50	44	50	48			
28th	30	49	43	41	55	57	35	37	39			
29th	36	43	43	41	55	57						
30th	34	50	50	42	46	48						
31st	34	50	50	44	46	48						

The hygrometric aspect of this climate seems to me to be completely misunderstood. In his well-known work on Pau, Sir Alexander Taylor makes the assertion, which has so often been copied by other writers as to be now regarded as an established fact, that "there is a peculiar absence of free communicable humidity in the atmosphere at Pau."*

If we test this statement by comparing the climate of Pau with some place where there is no "peculiar absence of free communicable humidity in the atmosphere," such as Kew, near London, for instance, we shall find little proof of this boasted superiority; thus during the month of November, in the first winter that I visited Pau, the mean moisture of the air denoted by the hygrometer was ·82 at Pau, compared with ·85 at Kew; and in December the mean moisture of the air at Pau was ·83, which was exactly the same as at Kew.

During my two visits to Pau, in October and January, there was a continual combination of cold and fogs in the mornings and evenings. In January the thermometer was frequently below freezing point during the night, and there were few days on which some rain did not fall.

It is certainly true, however, that there is here a peculiar stillness of the atmosphere, and generally a great freedom from harsh and violent winds, which is especially obvious in spring. This, though in some respects an advantage, is not an unqualified one, the stagnation of the air giving rise to a weak and languid circulation, generally accompanied by a corresponding state of mental and physical relaxation and inertia. And it has been remarked that the Bearnaise differ from the French in general, in being of less excitable disposition and more placid temperament. From this sedative action it has been argued that the climate should prove useful in chronic diseases attended by much vascular excitement, and indeed I was told by some of the English residents in Pau, that they had noticed a considerable reduction in the frequency of the pulse on their first arrival. I

* "The Curative Influence of the Climate of Pau and the Mineral Waters of the Pyrenees on Disease," by Alexander Taylor, M.D., p. 42.

may remark that this effect was not produced on those whom I had an opportunity of observing, but, on the contrary, the opposite symptoms followed the arrival in Pau of two persons under my immediate notice. In both cases the acceleration of the pulse was caused by the cold, damp, and ungenial climate of Pau aggravating the ailments of these individuals.

Rheumatism is a common complaint here, and a strong presumption is thus afforded that the climate is unsuited for persons suffering from that affection. Sir Alexander Taylor admits that bronchitis is not unfrequent in Pau during the winter and spring, but he considers that it is of a less severe character than in this country.

From the foregoing account of this climate and its effects I draw the conclusion, that the variable temperature of Pau renders that town unsuited for the winter residence of consumptive patients, in whom the disease has progressed to its second or third stage. It would be a bad abode in most cases of chronic bronchitis, and, according to the testimony of an eminent resident physician, peculiarly unsuitable in cases of fatty degeneration and certain other chronic diseases of the heart.

On the other hand, this climate may occasionally act beneficially in some cases of incipient phthisis, when a pure mountain air is required; and perhaps also in some instances of that disease in its first stage. But even in such cases I think that a more bracing and more equable climate is generally required.

With regard to asthmatic patients I must, however, modify what I said in my first work on this subject, as further experience has convinced me that in many cases of spasmodic asthma, Pau agrees better than any other climate.

Arcachon has of late years come into vogue as a spring and winter residence for pulmonary invalids, and there-

fore on my return from my last visit to Pau I made a short
stay in this village. Arcachon lies an hour's journey by railway to the south-west of Bordeaux, being situated in a pine
forest, and on the shore of an immense lagoon, which opens
by a narrow channel into the Bay of Biscay.

As Arcachon is chiefly known as a health resort for British
invalids, in consequence of Sir Dominic Corrgian's recommendation of it in a presidential address delivered some years ago,
and as this is now out of print, and contains the best account
yet published of Arcachon, I shall here summarise that
eminent physician's graphic description of this health resort.
According to Sir Dominic Corrigan, Arcachon owes its existence to the commander of a merchant vessel, who, in 1826,
established an hotel in this place. " Adopting the style to
which he had been accustomed in India, he built his first
establishment of wood after the model of a ' bungalow '—a
one-storied house with a verandah running round it, into which
all the rooms open. The model has been universally adopted,
and the watering-place of Arcachon now consists of hundreds
of such isolated houses, with magnolias, oleanders, and orange
trees around them, giving the whole place the picturesque
appearance of clusters of Indian bungalows in an American
pine clearing. In 1844 the opening of a railway gave a fresh
impetus to its progress, and Arcachon is now, apparently and
deservedly, one of the most favourite spring residences on the
south-west coast of France for invalids labouring under some
forms of pulmonary affections. I have obtained some short
notes of the temperature from Dr Hameau, the resident physician, and he assures me that during the last winter—which had
been unusually severe—the thermometer (Fahrenheit) only
fell four times in December, and three times in February to
freezing point, and this was on the beach. In January and
February, it appears from his tables that the temperature in

the forest is usually abou t44° to 50° of Fahrenheit. . . .
Arcachon is free from the vicinity of any high mountain
ranges that might pour down upon it cold, dry, and harsh air,
while it is sheltered by sandhills of moderate elevation ; and it
presents with this the additional advantage of sea air without
the violence of sea gales. There is another peculiarity to which
much of the salubrity of the air of Arcachon, as a residence,
is attributed—for it has a high local repute in pulmonary
affections—and that is the great belt of pine forest which
extends for many miles around it. The whole air is perceptibly
impregnated with the balsamic odour of turpentine, and we
know that the balsams and turpentines in vapour are remedial
agents of much power in bronchial affections." *

On the same coast, and also within the limits of the
same vast sandy plain in which Arcachon is situated, is
another locality which has still more recently been recommended as a winter resort, namely—Biarritz. Since I first
visited the district of the Landes the population of Biarritz
has more than trebled ; and, chiefly owing to the preference
given to it by the ex-Imperial family of France, this town has
been metamorphosed from a mere fishing village into one of
the most fashionable sea-side watering-places in Europe. It
is unnecessary to describe a locality so well known and so
accessible, being within five hours' journey by train from
Bordeaux, and only six miles from Bayonne. Nor would I have
made any reference to this town were it not that, not satisfied
with the repute which Biarritz has justly acquired as the first
amongst the sea-side resorts of France, an attempt has been
made by some writers to take advantage of the prevailing
opinion in favour of tonic winter climates for pulmonary invalids, and to write Biarritz, as other places have been, and with

* Sir D. Corrigan, " Introductory Address Medical Society College of
Physicians," p. 12.

even less justice written, into value as a fitting winter residence for consumptive patients. But notwithstanding its
unquestionable advantages of facility of access, beauty of
situation, and excellence of accommodation and living, still
the position of this town, which is fully exposed to violent
Atlantic storms and consequent frequent and sudden changes
of temperature during the winter and spring seasons, the prevalence of humid south-westerly winds, and the large number
of wet days, amounting on an average to one hundred and
twenty annually, obviously render Biarritz unsuitable as a
winter abode for pulmonary invalids.

MONTPELLIER. 79

CHAPTER XI.

MONTPELLIER.

TOWARDS the close of the last century Montpellier was the
best known and most frequented health resort on the Continent. But its reputation in this way has long been eclipsed
by other localities now in fashion.

This place is within sixteen hours' journey of Paris by the
Marseilles line, a branch of which, from Tarascon, passes the
town.

Montpellier is situated on a hill in the centre of an extensive
sandy plain, evidently once the bed of the sea, from which it
is now six miles distant. The streets are narrow and very steep,
the principal leading up to the "Place du Peyrou," from which
the tourist may enjoy one of the most beautiful views in the
south of Europe.

The town contains little to attract the notice of the visitor,
with the exception of the Musée Fabre, the Jardin des Plants,
and the ancient building occupied by the Faculty of Medicine,
amongst whose students I was once enrolled, and might therefore be suspected of partiality if I were to say more than that,
from its foundation in the tenth century down to the present
day, this school has maintained a reputation unsurpassed by
any of the larger and more modern centres of medical education. The classes are not very largely attended, however, and

the number of those who are here invested in the robe of Rabelais is now comparatively small.

In the vicinity of Montpellier the line of coast from the embouchure of the Rhone as far as Cette is bordered by lakes and marshes. The ground on which the town stands is of a loose gravelly formation, and in the neighbourhood a considerable amount of mercury is found in small cylindrical veins in the soil, and the peasants attribute the mortality amongst cattle, in certain localities, to mercurial exhalations from the earth.

The winter climate of Montpellier, although warm, is extremely variable. The mean annual temperature is 56°, which, compared with Malaga and with London, is 9° lower than the former and 6° higher than the latter city. As a comparison of this kind seems to me the best way of understanding the value of meteorological observations, I shall also apply it to the seasons. Thus the mean temperature of winter at Montpellier is 41°, or 14° lower than Malaga and 2° higher than London; the mean temperature of spring is 55°, or 13° lower than Malaga and 7° higher than London. The hottest month is July, and the coldest is January, and the mean daily range of temperature is 12°.

Within the last sixty years the climate of Montpellier has become considerably drier than it was. In the latter part of the 18th century the annual rainfall was 778 millimetres, and in the first half of the present century it only amounted to 662 millimetres, being a decrease of 116 millimetres a year within this period. The annual number of rainy days has fallen from 179 to 58 within the same time. The maximum number of wet days occurs in October, and the minimum in February.

The prevailing winds in Montpellier are, first, the northwest or "Magistráou." This blows for about 78 days annually, and during its prevalence the town is healthier than at any other time. The north wind or "Tramontana" is felt about 74 days in the year, and in winter, when it is termed the "Bise," is cold and harsh, but in summer is hot and parched. The east is a rainy wind, occurring on about 60 days annually, and crossing the marshes of Aiguesmorts, arrives at Montpellier pregnant with malaria. Southerly winds generally prevail in summer, and are cool and humid.

Although snow is rarely seen here, hard frosts are common during the winter, and violent hailstorms recur periodically every few years. Fogs are seldom observed, but the dew-fall is very considerable, so as to render it essential for invalids to remain within doors after sunset.

The atmospheric constitution of Montpellier is characterised by dryness and moderate warmth, accompanied by a tendency to great and sudden alterations of temperature, and the occasional prevalence of strong winds. It is manifest that a combination such as this cannot agree in cases of consumption.

In his excellent work, Dr Rodriguez has shown that "pulmonary consumption has increased in Montpellier since the close of the last century; it is often hereditary, and its exciting cause is generally ascribed to catarrh." "It is not uncommon," he tells us, "to meet with acute inflammatory phthisis or galloping consumption, especially in young girls."[*] Montpellier cannot therefore be recommended as a desirable winter resort for those suffering from phthisis, although even in that disease I am inclined to believe that this climate is as good as that of some other places recently brought into vogue as health resorts.

The cases which I think most likely to be benefited by the climate of Montpellier are cases of humoral asthma, and chronic laryngeal and bronchitic affections, attended with profuse

* "Clinique Medical de Montpellier," par le Dr Rodriguez, p. 382.

expectoration, and not accompanied with much irritation. These may in some cases improve rapidly under the influence of this climate, provided great care be taken to avoid exposure to the *bise* and *marin* winds.

The climate of Montpellier may be also tried in some forms of dyspepsia, climacteric disease, and hypochondriasis, attended with a slow circulation, and where the feelings are not morbidly acute.

CHAPTER XII.

HYÈRES AND CANNES.

FROM Montpellier to Hyères, there is a wide transition of climate, although both towns are within the ancient limits of " Fair Provence." For brilliant as is generally the sun, cloudless the sky, and serene the atmosphere of this entire region, yet the extreme inconstancy of the temperature renders the greater part of it unsuitable for the residence of pulmonary invalids. So changeable indeed is the climate, that Louis XIV., who certainly knew something about ladies, if nothing about climates, very happily compared it to "*une coquette parfumée dont il fallait se méfier.*"

A long narrow strip of land extends along the Mediterranean from Toulon to Nice, and thence is continuous with the Riviera del Ponente. This is included between the secondary chain of the Maritime Alps and the sea, and being protected by these mountains from all cold winds, and especially from the " Mistral," is thus blessed with a climate very different from that of Provence in general. Enjoying the advantages of this situation in a more than ordinary degree is the locality now to be described.

The town of Hyères is situated in the department of the Var, near the western boundary of the Gulf of Lyons, about two miles from the shore, and is connected with Paris by the railway to Toulon, from which it is nearly twelve miles distant on the road to St Tropez. The town is small, having a population of some 10,000 inhabitants. The older portion is built on the side of a precipitous hill.

Foreign invalids for the most part inhabit the modern part of the town about the Place de Palmiers, Place Royal, and Place de la Rade, where the best lodging-houses and hotels are situated. There is a " Cercle " or club, and a lending library, both well supplied with English books and papers. An English Protestant church has also been erected in the Place de Palmiers. English physicians reside in the neighbourhood, and amongst these I may be pardoned for mentioning the name of my friend Dr Griffiths, who has been for many years in extensive practice in Hyères, and to whom I am indebted for valuable information respecting the climate. Living is cheap and good; and all the necessaries, and most of the commodities, a sick man requires, are to be easily procured here.

The climate of Hyères is warmer and more equable in winter than that of Nice; it is also stated to be drier, but at the same time less exciting than that town. The average number of rainy days in Hyères being only 40, while in the latter locality it amounts to 60 days yearly ; and the number of inches of rain annually is about 27, which is a little more than at Nice.

During the winter the prevalent winds are north, north-east, south, and south-east; and in spring they are thus arranged by M. Carrière, in the order of their frequency—east, south-east, and north-east.

Dr Griffiths, in a note appended to some valuable meteorological tables with which he favoured me, says,—" The rain, it will be observed, mostly falls at night, so that there is rarely a day when the invalid may not go out for some little time. The 'Mistral' (N.W. wind), as described in 'Murray,' is almost never seen in Hèyres; and when it does prevail, is not to be compared, for dust and discomfort, to the 'Terral' in Malaga."

Snow is seldom seen here, and according to Dr Gigot-Suard it only snows once every three years, and then, but for a very short time.*

Generally the climate may be said to be fitted for children or young persons of a lymphatic temperament, or of a scrofulous diathesis, either predisposed to consumption, or suffering from the first stage of that disease.

Bearing in mind that the atmosphere though drier, is at the same time less exciting than that of Nice, a residence in Hyères would be likely to prove useful in cases of chronic mucous discharges, chronic bronchitis, neuralgia, atonic dyspepsia, and some other diseases, in which the climate of Nice would be unduly tonic, or stimulating.

Cannes is situated at the extremity of the bay of the same name, on the road from Toulon to Nice, and is about twenty-one miles to the south-west of the latter town. The plain on which it stands is enclosed on the north and west by the Maritime Alps and the Estrelles, but on the east this mountain barrier is not complete; while towards the south the plain is open to the Mediterranean.

Opposite to the town are " les îsles de Lérins," of which the larger, St Marguerite, is famous for having been the scene of the long incarceration of the mysterious "Man in the Iron Mask," in the seventeenth century.

In the valley of Cannes, vegetation is peculiarly luxuriant. The natural beauty of the scenery is much increased by the soil being mainly devoted to the culture of odoriferous plants and flowers, from which perfumes are [here very extensively manufactured.

The town, though a thriving place, and possessing a

* Gigot-Suard, " Des Climates," p. 154.

population close on 5000 inhabitants, contains in itself but little to interest the casual visitor. The principal street, containing a few hotels and lodging-houses, is built along the high road to Italy, and is separated from the beach by a public promenade. On the hill above this is the ancient Church of "Nôtre Dame d'Esperance," a shrine renowned among the mariners of Provence.

On the same side are the environs, chiefly inhabited by foreign invalids, and also the Scottish church, above which is the magnificent castle of the Duke de Valambrosa.

The climate of Cannes is peculiarly equable as well as moderately warm in winter. The mean annual temperature is about 60°; and Dr Sève, of this town, gives as the result of the observations of fourteen years, the following table of the mean temperature of the seasons :—

Winter,	. . 50°	Summer,	. . 71°
Spring,	. . 62°	Autumn,	. . 55°

The prevalent wind at Cannes is the sea-breeze from the east and south-east. North and west winds are, to a certain extent, excluded by the maritime Alps on the one side, and the chain of the Estrelles on the other. Violent winds, and the "Mistral," are said to be less felt here than at Hyères.

The annual number of rainy days at Cannes is fifty-two, or twelve more than at Hyères, and eight less than at Nice. Whilst with regard to the actual amount of rain which falls, the proportion is reversed, as the annual rainfall at Cannes amounts to 25 inches, being 2 inches less than that of Hyères, and the same as Nice.

The exciting character of the climate of Cannes is, by native authors, attributed to a highly electrical condition of the atmosphere.

I have very little to add with respect to the sanative application of this climate; but its electrical condition, its equable warmth and dryness, and the stimulating properties of the atmosphere, all would indicate its fitness for scrofulous and lymphatic temperaments. It might probably be also resorted to in some few cases of consumption occurring in persons of these temperaments, and marked by symptoms of langour and debility; in similar cases of chronic bronchitis, and in certain low nervous affections.

CHAPTER XIII.

NICE.

As a sanatorium for foreign invalids, Nice possesses some
natural advantages, as well as many artificial ones.   Among
these may be included the many facilities for reaching it by
land and sea, the beauty of its situation, and the superiority of
the accommodation it affords; all of which are important
accessories to the purity of its atmosphere, and the serenity of
its sky.   So far back as the fourth century we find these latter
celebrated by Ausonius :—

> " Nicæa est Natale solum, clementia cœli
>   Mitis, ubi est riguæ larga indulgentia terræ,
>   Ver longum, brumæque breves, juga frondea subsunt."

Situated on the verge of a valley formed on three sides by
the Maritime Alps, and opening to the Mediterranean on
the south, Nice lies 170 miles from Marseilles, and 213 miles
from Genoa.   The mountains, which in great measure shelter
the valley of the Paillon from cold and harsh winds, also
serve to concentrate and radiate the solar rays on the town,
and in some degree account for the peculiar mildness of the
climate during the winter.

The town is built around the base 'of a lofty promontory
surmounted by the ruins of a fortress. To the east of this
rock is the port, and around its base, on that side, are congre-
gated the crowded, narrow streets of the old town.   The river

Paillon, which flows through the town, is bordered on its eastern quay by a handsome boulevard which connects the port with the new town. Opposite the Pont Neuf is the suburb of the Croix de Marbre, which is principally occupied by foreign invalids, of whom the majority are English or Americans. The hotels, of which the visitor has his choice of at least a score, are chiefly situated on the western side of the river, or in the Promenade des Anglais, and are generally better than in most French provincial towns, and also somewhat more expensive.

Next to the climate of a place, the *agrémens* or amusements it possesses are justly the most important consideration for the invalid visitor, who having abandoned his accustomed employment, and seldom bringing books with him, depends on them for the means of occupying his mind, and spending his time. These are better provided for in Nice than in any other southern city I am acquainted with, except Naples. Moreover, even in the depth of winter there are very few days in Nice on which some hours of sunshine and warmth do not occur, and, conjoined to the advantages of good roads, easy conveyances for reaching the many places of interest about, and the exquisite beauty of the scenery in the neighbourhood, afford the most delicate valetudinarian an opportunity and temptation for going out, and thus availing himself of the great accessories to health—pure air and moderate exercise.

The climate of Nice is essentially of a dry, warm, tonic, and exciting nature. According to the Chevalier Macario, whose calculations are founded on the observations of preceding writers for the last fifty years, the mean annual temperature of Nice is 60°; that of winter 48°, spring 55°, summer 71°, and autumn 62°. The temperature is very steady, the variations from month to month not exceeding from 2° to 3°.

There are, on an average, 229 bright, cloudless days annually

at Nice, 66 cloudy days, and about 60 rainy days, the average
rainfall amounting to 26 inches, of which one-half falls in
autumn, and one-fourth from October to February.

In winter the prevailing winds in the order of their frequency
are north, east, and south; and in spring east and south winds.
The northerly winds from the snow-covered Alps are attended
by cold, dry weather, and principally prevail in spring, at
which season, conjointly with the east or west winds, they
occasionally blow with great violence along the valley of the
Paillon, and those parts of the town that border on the river.

The "Mistral" is shut out to some extent by the mountains
between Fréju sand Cannes; still, a cold, dry, irritating wind
from the north-west occasionally occurs, but only lasts for a
few hours. The east winds in spring are the great drawback
to Nice, and during their prevalence no invalid should think
of leaving the house.

We have now to consider the practical application of the
foregoing observations. And on this subject we find the most
contradictory opinions expressed by various writers on the
climate. The majority of invalids who are sent to Nice suffer
from pulmonic disease of some kind, " *Voulez-vous savoir,*"
tersely asks Monsieur Champouillon, "*ce que deviennent les
tuberculeux à Nice?   Allez au cimètiere.*" Dr Fodéré of
Strasburg, who lived in Nice for several years, says—"Here
the disease is not chronic as in Switzerland, or in Alsace;
but I have very often seen it terminate within forty days,
and a physician of the country I have just named would be
astonished with the rapidity with which hæmoptysis sets in,
the tubercles suppurate, and the lungs are destroyed."*

Dr Farr says—" Independently of the 'Mistral,' the easterly
wind sets in with the first moon in March. I besought those

* " Voyage aux Alpes Maritimes, ou Histoire Naturelle, Du Pays de Nice."
p. 184.

whom I attended, and many whom I did not, to quit Nice before the birth of this fatal moon; but they heeded not my counsel, and thought that I had over-rated the danger. They remained, and the day after this easterly wind began; of the thirty, I only met one afterwards, and him I had often previously pronounced to have no disease of the lungs." *

A resident physician, Dr Wahu, states that—" It would be the greatest mistake to assert that Nice, on account of its climate and hygienic conditions, is the locality which agrees with phthisical patients."†

The late Sir James Clark says—" Indeed, sending patients labouring under confirmed phthisis to Nice will, in a great majority of cases, prove more frequently injurious than beneficial."‡

To the preceding citations against the climate of Nice I had intended to add others equally strong in its favour, from the works of resident physicians, who either entirely deny, or explain away, the foregoing conclusions.

Amongst these I had taken extracts from Richelmi,§ Roubaudi,|| Chevalier Macario,¶ Dr Edwin Lee,** and some others. But it has been deemed advisable to leave out these quotations, since they only serve to illustrate the difficulty and uncertainty of the study of climate.

I have already characterised the climate of Nice as being moderately warm, dry, and somewhat exciting, therefore it may be expected to agree best with those forms of disease in which debility and langour, attended by profuse expectoration, or some other exhausting discharge, are prominent symptoms. A

---

* Dr W. Farr, " On the Climate of Nice," p. 16.
† Le Dr A. Wahu, " Conseiller Medical de l'Etranger à Nice," p. 17.
‡ " The Influence of Climate," &c., p. 123.
§ " Essai sur les Agréments et sur la Salubrité du Climat de Nice."
|| " Nice et ses Environs."
¶ " De l'Influence Medicatrice du Climat de Nice."
** " Nice and its Climate."

residence in Nice often proves eminently serviceable to children of a weak and relaxed constitution, predisposed to any of the forms of scrofula, and therefore to consumption. It is also resorted to, though with less marked benefit, in cases of consumption, either in its premonitory state, or at any rate before it has passed into the second stage, occurring in a torpid or leuco-phlegmatic habit of body, and not accompanied by symptoms of active inflammation.

Patients suffering from chronic bronchitis and senile winter cough, often derive singular benefit from the climate of Nice, the effect of which seems to consist in a diminution of the excessive mucous discharge, that is often the chief cause of the patient's weakness. And in cases of humoral asthma I have hesitated between recommending Nice or Malaga.

Cases of chronic rheumatism occurring in persons of scrofulous habit occasionally derive advantage from a residence in this town; but it must be recollected that the climate is not exempt from great and sudden atmospheric vicissitudes, and that rheumatic patients must be very cautious to avoid exposure to these. Change of climate undoubtedly exercises a wonderful control over the manifestations of gout, and in most cases a dry, warm climate, such as this, is calculated to be useful.

The hills of Cimiès, and Carabacel, to 'the north-east of the town, contain some of the best residences in the vicinity of Nice, and the climate would render them all that could be desired, were it not that the deficiency of good drinking water is a very considerable drawback. To the north are Le Ray and St Barthélemy, which have a similar aspect to Cimiès, and are suited for the same class of patients.

CHAPTER XIV.

MENTONE AND THE RIVIERA.

WITHIN an hour's journey by railway from Nice is a locality
which has been very assiduously and ably written into vogue
as a winter resort for pulmonary invalids, and has rapidly risen
from an obscure village to the position of one of the best known
and most fashionable health resorts of southern Europe. The
reputation which Mentone has thus acquired is chiefly due to
Dr J. H. Bennet who, having here regained his own health,
impaired by arduous professional labour in London, concluded
that the climate which agreed so well with himself must needs
suit other invalids, and published a work which, from its first
edition to the last expanded treatise he has recently brought
out on the same subject, has throughout one predominant
theme, namely, the reiteration of the asserted superiority
of the Mentonian climate.

Mentone, or Menton as the French call it, is an ancient
Italian-looking town, charmingly situated on the sea-shore,
about twenty miles east of Nice, with which it is connected
by railway. The town itself consists of a long, straggling
street running parallel to the shore, and of a few lanes which
ascend the hill, whilst outside the town are scattered on all
sides a considerable number of large hotels and villa residences
which are chiefly occupied by English, American, and other
foreign invalid visitors.

To meet the wants of so large a transitory population

English shops have been opened, English doctors have estab-lished themselves, and a club with assembly rooms, concert hall, &c., set up. The principal hotels are tolerably good, and the charges are about the same as at Nice.

Before Dr Bennet became the panegyrist of the Mentonian climate, this place had been favourably noticed as a health resort in M. Carrière's well-known work on the Italian climates, and long previously its supposed advantages had been pointed out by Dr Fodéré of Strasburg. The advantages attributed by those and other writers to the climate of Mentone are, a warm and equable temperature, exemption from harsh and cold winds, and particularly the "Mistral," a somewhat sedative atmosphere, facility of approach, and convenient residences suitable for invalids.

According to M. Bréa, the mean annual temperature of Mentone is 60°; the maximum observed was 80° in August, and the minimum 32° in January.

Mean Temperature of the Months.

January,	. 48°	May,	. 63°	September,	. 69°
February,	. 48	June,	. 70	October,	. 64
March,	. 52	July,	. 75	November,	. 54
April,	. 57	August,	. 75	December,	. 49

Resumé of the state of the Weather during 3645 days (M. Bréa).

Fine,	. . . . . .	2140
Partly clear, partly cloudy,	. .	457
Cloudy,	. . . . .	248
Rainy,	. . . . .	800
Total,	. . .	3645 *

The author of a recent English work on Mentone, tells us, that the "Mistral" wind is unknown in that favoured climate,

* "Tableaux Synoptiques des observations météorologiques faites à Menton (*Alpes Maritimes*), par M. de Bréa, Sous-Intendant Militaire en retraite."

but at the time of my visit to Mentone in March, I was informed that the "Mistral" had been raging for some days previously, and my informant was much amused at my incredulity on the subject.

Nor, considering the position of this village, is it possible that it can be exempt from strong and cold winds; situated in front of Corsica, whose mountains are covered with snow during the winter, it cannot escape the ill effects of that proximity.

Dr Bennet, in the first edition of his work on Mentone, says:—"To live at Mentone is really like living on ship board, for, as already stated, the greater part of the inhabited tract is a mere ledge at the foot of the mountains. There are very few houses inland."* In this passage Dr Bennet has, however unintentionally, made use of a strong argument against Mentone as a residence for invalids; for, as Sir Dominic Corrigan well observed in the monograph I have already quoted :—"No locality that is small in extent, no matter how favourable it may seem, is desirable; for if a turn round a hill, or a different aspect at a short distance, give a considerable change of temperature, the locality is unsuitable, both on this account and because the resident is there confined to too small a space, and body and mind suffer."

Dr Edwin Lee in his "Notice of Menton," says—"Inflammation of the lungs, pleura, and air-passages, not unfrequently terminating after repeated attacks, or from debility of constitution in phthisis, is of frequent occurrence amongst the inhabitants, especially of the lower class, who, being badly fed and clothed, are sooner affected by atmospherical changes. Consumption, likewise, sometimes occurs, irrespective of any such acute attacks from anti-hygienic causes analagous to those I have specified in the account of Nice. The enervating

* "Mentone and the Riviera," by J. H. Bennet, M.D., p. 61.

influence of the climate of Menton on those destined to pass the whole year there, produces, however, some diseases which are comparatively infrequent at Nice." *

For my own part, I need hardly add that I do not share the opinion of those who consider Mentone the Utopia for invalids generally, and the climate *par excellence* for pulmonary sufferers in particular.

Less than an hour's journey by the railway beyond Mentone, and just within the Italian frontier, is the rival health resort of SAN REMO, which, being some fifteen miles further from the gambling tables of Monaco, which are the real though unacknowledged centre of attraction for some tourists to the Riviera del Ponente, is as yet less crowded by foreign valetudinarians. However, it already attracts a considerable number of winter residents; and having excellent hotel and lodging-house accommodation, and enjoying a climate very similar in all respects to that of Mentone, may be resorted to, or should be avoided by, the same classes of travellers for health. In fact the whole Riviera from Nice to Spezia is studded over with small towns, such as Bordighera, Savona, Voltri, Nervi, and many others, which, having the same advantages of picturesque position and mildness of climate as Mentone, require only improved accommodation and well written and roseate-coloured medico-descriptive eulogy from some resident physician, to attract as many of the wealthy valetudinarian population of these islands, who are always seeking a new health resort, in which to escape the leaden sky and ungenial atmosphere of our winter climate.

* "A Notice of Menton, supplementary to Nice and its Climate," by Edwin Lee, M.D., p. 21.

CHAPTER XV.

PISA AND ROME.

THE prevailing opinion in favour of dry or tonic climates as health resorts for pulmonary invalids, well founded as it is in the majority of cases of phthisis, has led to the comparative abandonment of sanatoriums such as Pisa, which, though not belonging to that category, are valuable in other chronic diseases.

It would be superfluous in a work on climatology to enter into any general description of a city so familiarly known as this. But it will be necessary, however, to indicate briefly the topographical peculiarities which influence its climate.

Pisa, then, is situated near the maritime extremity of the rich alluvial valley of the Arno, thirteen miles north of Leghorn and forty-five miles west of Florence, with both of which it is connected by railway.

The surrounding country spreads out towards the north-west into an extensive plain extending to the sea, near which the soil becomes swampy, and forms the well-known unhealthy fens of the Maremma Pisana, and Maremma Volterrana; whilst on the opposite side of the town the chains of the Apennines extend.

The city is encompassed by lofty walls nearly six miles in circumference, which contain within them a population of about 22,000 inhabitants. Some writers attribute the peculiar stillness of the atmosphere of Pisa to these walls, which they suppose act as barriers against the force of the winds.

Pisa is divided into two very different climates by the Arno, which is here much wider, and if possible still more turbid, than at Florence. In its course through the town the river describes a curve, the convexity of which is directed to the north, and the Lungo l'Arno to the south. This disposition serves to concentrate the solar rays on the former side, to the disadvantage of the other, and therefore on the Lungo l'Arno are situated the principal hotels and houses inhabited by foreigners, and no pulmonary invalid should think of residing on the opposite side, where the houses are damp and perceptibly colder.

To the visitor newly arrived from Paris or Naples, where pleasure and gaiety seem to be the chief business of life, Pisa presents a dreary and desolate aspect, and the traveller, who, recollecting the story of her long enduring greatness and rapid decadence, now visits her, may here learn a wholesome lesson on the instability of national prosperity.

The climate of Pisa is essentially humid and warm, approaching in these respects closer to that of Madeira than any other part of Europe. The humidity of the atmosphere, characteristic of Pisa, is a necessary consequence of the position of the town, and of the direction of the prevailing winds, which all pass over large surfaces more or less completely covered by water.

The mean annual temperature of Pisa is 59°; and that of the seasons is—winter 44°; spring 57°; summer 73°; and autumn 62°.

The prevailing winds are those from the south and south west, as the town is protected from the north-west to the south-east by the high mountains in that direction. The south wind is warm and moist in winter, and is favourable to

the class of invalids for whom this climate is adapted. But in
summer it is dreaded as the hot "Scirocco," and brings with it
the seeds of disease, from its passage over the stagnant
Maremma.

Pisa possesses an atmosphere eminently sedative, and anti-
phlogistic. Irritation of the pulmonary mucous membrane,
accompanied by a hard, dry cough, is assuaged, expectoration
is increased, and the pulse becomes slower and softer.

Such effects point out the power of the climate in allaying
dry chronic bronchitis, especially in the aged, and at the same
time warn us against sending consumptive patients generally,
and more especially in the second stage, to this place, where
the relaxing influence of the warm, moist atmosphere, would
soon manifest itself in the augmented expectoration, diminished
strength, and accompanying night perspirations, or diarrhœa,
which would in all probability hasten the fate of the sufferer.

In many cases of consumption it has been observed, that
even a short residence in the warm, humid, and relaxing atmosphere of Pisa has been followed by severe hæmoptysis.

For children also the Pisan climate is generally too relaxing,
and the body does not here acquire that healthy tone produced
by a more bracing atmosphere.

Rome belongs to the same class of climates as Pisa and
Madeira, being humid and warm in winter and spring, but less
equable than either of these health resorts. The late Sir
James Clark, however, considered that in range of temperature
Rome has the advantage of Pisa; and from his own experience
arrived at the conclusion that "the climate of Rome in regard
to its physical qualities is altogether the best of any in
Italy."* More recent writers, however, do not agree in this
opinion.

The mean annual temperature of Rome is 60°, or 8° lower

* Sir James Clark "On the Influence of Climate," &c. p. 143.

than Madeira, and the same as that of Pisa. The mean temperature of winter is 49°, or 11° lower than Madeira and 2°
higher than Pisa ; the mean temperature of spring is 58°, or 4°
lower than Madeira and 1° higher than Pisa ; in summer the
mean temperature of Rome is 74°, or 3° higher than Madeira
and 3° lower than Pisa ; and in autumn the mean temperature of Rome is 62°, or 4° lower than Madeira and 1° higher
than Pisa.

The annual number of rainy days in Rome is 117. During
the winter and spring southerly winds are prevalent; but
before and after sun-set a cold, northerly breeze is commonly
observed at these seasons   The atmosphere of Rome is generally peculiarly still during the winter. But this stillness is
occasionally interrupted by a strong, harsh, north-east wind,
known as the " Tramontana," the " Aquilo " of the Ancients,
the ill effects of which were graphically described by Celsus.
This wind, like the "Terral" of Malaga, seldom lasts longer
than three days, during which no consumptive or bronchitic
patient should venture out of doors. Towards the middle of
March sharp easterly winds generally set in, and, being
accompanied with a hot sun and cloudless sky, are attended
with all the ill effects of a great and sudden change of
temperature, to which no pulmonary patient should be exposed, but should leave before this time.

The prevailing diseases here are nervous and spasmodic
affections and cerebral diseases, particularly apoplexy, at all
seasons. In winter pneumonia and pleuritis are common, and
in summer and autumn febrile disorders predominate. For
this reason Rome should be avoided by valetudinarians from
June to October, as malarial fevers are then endemic.

It is almost useless to reiterate the hackneyed but well-
founded warning to invalid visitors in Rome as to the danger
they so recklessly expose themselves to in passing from the

warm and sunny streets into the chilly galleries and palaces
which invite their inspection on every side. The only safe resort for the valetudinarian pilgrim to Rome is that mighty
shrine—

> " —the vast and wondrous dome,
> To which Diana's marvel was a cell."

It is a remarkable fact that, notwithstanding its vast ex-
panse and complete shade, the atmosphere of St Peter's is
equally genial and uniform at all seasons, and may be visited
all times with safety by any invalid.

In cases of chronic bronchitis and winter cough, attended
with great irritability of the respiratory organs, constant hard
teasing cough, and little expectoration, as well as in similar
cases of chronic laryngeal disease and spasmodic asthma; and
also in chronic rheumatism, rheumatic gout, and gouty
bronchitis, there is hardly any European climate so suitable
for a winter resort as this. But in consumption the sanative
influence of the Roman climate must be admitted to be limited
to a comparatively small number of cases, and I need not here
repeat what I have said on this point in the first chapter of
this work.

CHAPTER XVI.

NAPLES.

NAPLES has been long esteemed as the *beau idéal* of all that is beautiful in situation, or delightful in climate. " Vedi Napoli," says her own proverb, " è poi muori,"—

> " And sooth to say who sees her will retain,
> In his mind's eye a gorgeous soil and clime,
> The last to vanish with the lapse of time." *

The easiest and most advantageous route by which the invalid traveller can reach Naples is direct from Liverpool, London, or Glasgow, from each of which ports there are regular fornightly sailings by large and commodious steamers. Those, however, who dread crossing the Bay of Biscay may still reach Naples, with comparatively little fatigue, *via* Marseilles, in four days from London. The Mont Cenis route, though occupying only a little more than half that time, is not, I think, generally suitable for invalid travellers.

As seen from the bay, Naples has been said to form a vast triangle, the base of which rests on the shore, and the apex is placed at Capo di Monte. The city is divided into nearly equal parts by the Via Nazionale, as the Strada Toledo is now called, and its continuation, the Strada Nuova di Capodimonte, which intersects it for close on three miles. The lower part,

---

* Dr William Beattie, "The Pilgrim in Italy," Cant. 2d.

formed by small narrow streets, extends towards the eastern side of the bay, and on the other side, which lies above, and parallel to the shore, towards Posilipo, contains the best quarters, the Chiatamone, Santa Lucia, and the Chiaja. The hotels, which are principally patronised by our compatriots, are those on the Chiaja, which, however, are expensive, and, as far as my experience goes, not over comfortable. Therefore I should recommend those who purpose spending the winter in Naples to procure lodgings in a good situation, or to locate himself in one of the numerous Pensions which may be found along the Riviere di Chiaja, the Corso Vittoria Emanuele, and almost all the principal streets. Before deciding on a residence, however, the invalid visitor should consult some local practitioner, since so great are the differences in aspect between the different parts of the city, that Naples has been said to possess two perfectly distinct climates, of which the western, or more elevated part is exposed to the harsh cold north-west wind, while the eastern side, or that lying along the plain, is protected from the north, but fully exposed to the relaxing, moist, and warm southerly winds.

In no respect does Naples contrast so favourably with every other place frequented by invalids, as in the number and variety of resources for passing time agreeably which it possesses, and the opportunities which it thus affords of withdrawing the attention of the valetudinarian from his ailments. Beyond the city walls extends on every side a country unequalled for the beauty of its scenery, as well as in the historic interest attaching to each spot. And within its limits, nothing has been left undone by man, to adapt to his tastes and pleasures the paradise which nature has given him to inhabit. The most magnificent churches; the most extensive museums, numerous palaces, libraries,—the envy of the learned of every land,—the most beautiful gardens; and theatres, such as the San Carlo, the most splendid in Europe, being here collected together.

The climate of Naples is dry and warm in winter, but at the same time is most changeable and uncertain at all seasons. The mean annual temperature is 61°, that of winter 48°, spring 58°, summer 70°, and autumn 64°. Compared with Malaga the mean temperature of Naples is 4° lower; that of winter 7° lower, spring 10° lower, summer 8° lower, and autumn 4° higher than Malaga.

According to the observations of M. De Renzi, extending over twenty years, the mean annual rainfall is about 29 inches,* of which the largest proportion falls in autumn, and when westerly winds suddenly succeed to a south wind, as the atmosphere is then saturated with vapour, which is condensed by the cool west wind, and falls as rain. The average number of rainy days is about 100, of cloudy days 120, and of fine days about 145.

The position of the hills, which encircle the city, protects Naples to some extent from northerly winds. The prevailing winds are,—the south-west or " Libeccio," which is often accompanied by severe gales and cloudy weather; the south, or " Scirocco," attended by an oppressive heat, causing great depression of mind and lassitude of body; the west, or " Ponente," during which the sky is generally serene, and the temperature warm in winter and cool in summer; and the north-west, or "Maestro," which resembles the " Mistral " of Provence.

No pulmonary patient should remain in Naples during the spring, whilst the cold and ungenial easterly and " maestro " winds are prevalent, occasioning rheumatism and pleurisy. The best time for such patients to inhabit Naples is from the end of autumn until Christmas. For those, who

* " Topografia di Napoli," p. 57.

NAPLES.   105

from choice or necessity, remain, as I have done, in Naples in
summer, the opposite shores of the bay afford a wide selection
of cool and agreeable retreats from the heated and not very
odoriferous atmosphere of the port and adjacent parts of the
city.

The frequency of sudden changes in the temperature,
hygrometrical condition, force, direction, and pressure of the
air, as well as the volcanic agencies in constant operation in
the vicinity of the town, all contribute to produce an atmosphere highly charged with electricity. This manifests itself
by the common occurrence of atmospheric perturbations, such
as thunder storms, and in the nervous excitement and uncomfortable sensations which all, and especially invalids, complain
of at such times.

Amongst endemic diseases, rheumatism and catarrh are
common. Strangers are very liable to pneumonia and pleurisy;
and phthisis undoubtedly occasions a large proportion of the
total mortality.

With regard to the therapeutic application of the climate of
Naples, it must be admitted that this city possesses great
advantages for hypochondriacal and melancholic patients;
for most cases of dyspepsia; for persons suffering from
climacteric disease, or from impaired health, the consequence
of long residence in tropical climates. In such cases the
mental impressions produced by the cloudless sky and
beautiful scenery of Naples, will seldom fail to co-operate
with the stimulating and electrical atmosphere in arousing the
nervous energies, and exciting the action of the heart, both of
which in these cases are generally weak and torpid.

For pulmonary patients, however, and especially for consumptive invalids, Naples is not a suitable health resort.

CHAPTER XVII.

FROM Naples the voyage to Palermo may be accomplished in ten hours, by steamers starting every other day.

Palermo is situated on the northern coast of the island, and is built on a well-sheltered bay, extending from the base of the rocky " Monte Pellegrino," on the west, to a slope of richly cultivated gardens, which gradually ascend to Cape Zaffarano on the opposite side. The plain, aptly termed by the natives the " La Conca d'Oro " or Golden Shell, is defended from all harsh and cold winds by the mountains around the city, and is perhaps the most beautiful, as it certainly is one of the most highly cultivated, valleys in Europe, not excepting even the Vega of Malaga.

From the roadstead the city seems crowded with churches and palaces, whose tapering steeples and swelling domes rise above the massive fortifications that enclose the town. Along the beach extends the Marina, which is little inferior to the Chiaja of Naples. The streets are clean and regularly paved with lava, but offer to the stranger a picture of the close relationship of splendour and poverty, the immense palaces which adorn many of them being occupied on the ground floor by poor shops and cobbler's stalls. The principal thorough-fares, however, such as the Cassaro, or Strada Toledo and Strada Nuova, each over a mile in length, are handsomely and regularly built, and intersect the town, forming by their junction a fine square, the Piazza Villena.

In no part of the world does cultivation appear to be carried to a greater extent than here, and nowhere does the fertility of the earth more abundantly repay the toil expended on it. The usual produce of wheat is from ten to sixteen fold, and in favourable years even thirty fold. Grapes are no less abundant, there being some twenty different species in the island, of which the most esteemed are the Muscatel and Canicula. The orange and lemon trees flourish in every valley, and the stunted olive grows almost wildly. Among the other productions, figs, almonds, saffron, and pistacio nuts may be instanced in proof of the mildness of the climate.

The mean annual rainfall in Palermo averages about 25 inches, or 3 inches more than London. The mean annual temperature is 62°, or 12° higher than London, 1° higher than Naples, 3° lower than Malaga, and 8° higher than Pau. The thermometer in the hottest days seldom rises as high as 90° or 92° (in shade), or falls lower than 36° even in the depth of winter; and Captain Smith tells us that in one year "there were 121 overcast and cloudy days, on 85 of which rain fell; 36 misty days; and 159 fine bright days."*

The prevailing winds are the northerly and westerly, both considered dry and healthy, and a variety of the "Mistral" called by the Palermitans "Mamatili," and enjoyed by them as an agreeable sea-breeze. Occasionally, in summer, the "Scirocco" prevails, and is more severely felt in Palermo than in any other part of Southern Europe.

Fogs are said to be of very rare occurrence here, but I have seen as dense a fog as ever I witnessed off the Sicilian coast, so thick, indeed, was it, that our captain was obliged to lie

* "A Memoir descriptive of Sicily and the Islands," &c., by Captain Smith, R.N. p. 4.

to for some hours, to the great alarm of the Neapolitan passengers. These fogs have not, however, the same depressing
and injurious effects as ours, but are rather a thick warm
haze, possessing singular refractive powers, magnifying and
altering surrounding objects to a surprising extent, and
occasionally giving rise to the celebrated spectral illusion of
the *Fata Morgana.*

The nights are generally cold, and the quantity of dew
which falls after sunset is sufficient to compensate for the
want of rain; this should warn the invalid to avoid exposure
to the night air.

Whether the climate of Palermo be suited or not for consumptive patients, is a disputed question. But it cannot, I
think, be doubted, that during the winter it is in some
respects superior to Naples for such invalids. Towards the
commencement of spring, however, cold and harsh winds
predominate; and I know of two cases within my own
practice that proved fatal here at this season, in both of
which the sufferers had improved in health during the
preceding winter at Malaga.

I would recommend the climate of Palermo to the majority
of dyspeptic and hypochondriacal patients, and, in fine, to
nearly the same class of invalids to whom I stated the climate
of Naples would probably be of service.

CHAPTER XVIII.

MALTA.

For the British valetudinarian in search of a southern health resort, Malta would at first sight appear to offer peculiar advantages. It is certainly further to the southward than any other part of Europe. It is easy of access from England ; the expenses of living are moderate ; and all the surroundings of the invalid in Malta are more English than in the other foreign winter resorts described in these pages. But whether the climate is as suitable for pulmonary invalids as some contend or not, is a question which the following details will enable the reader to judge for himself.

The island of Malta, which lies about sixty miles south of Cape Passaro in Sicily, and two hundred miles from the African coast, is close on sixty miles in circumference, and presents a rocky arid plain sloping from the south-west to the north-east. The geological formation of the island explains the almost tropical temperature which prevails here in summer, and the slight difference between the heat of the day and night at this season ; for the barren limestone rock absorbs the excessive solar heat during the day, and slowly radiates this during the night, thus equalising the temperature of the twenty-four hours.

Valetta is situated on the western side of an elevated Peninsula, Mount Xiberras, between the great harbour, " Di

Libera Pratica," the finest natural haven in Europe, and the small port of Marsamuscetta.

On landing we disembark at the famous *"Nix Mangiare"* stairs, so named from the unceasing cry of the importunate beggars, who here appeal to each new arrival; and passing through the well-stocked fruit market ascend the long stone stairs that lead to the Strada Reale, the principal street of the town. This runs through the whole length of the city from the fort at the entrance of the harbour, traversing the Piazza, where the palace of the Grand-Master, now the residence of the Governor, is situated, and terminating at the Porta Reale or gate on the inland side of the town. The houses in this, as in most of the other parts of the city, are handsomely and regularly built of cut stone, and the streets are kept strictly clean. The other streets either run parallel with the Strada Reale, or communicate with it by flights of steps once apostrophized by Lord Byron—

> " Adieu, ye cursed streets of stairs,
> How surely he who mounts them swears."

The many public edifices and monuments of the Knights of St John which Valetta possesses, the cathedral, public libraries, university, hospitals, and fortifications, together with the interesting excursions that may be made into the interior of the island,—for instance to Citta Vecchia, or to the Catacombs and Grotto of St Paul, near Rabbato ; and the short and delightful sail to Calypso's fabled island of Gozo ; are all to be included among the attractions of this town as a residence for dyspeptic and hypochondriacal sufferers.

Living is remarkably economical, particularly in the parts of the island furthest from Valetta, and though prices have risen considerably of late, yet still, house-rent and lodgings are very cheap, and even in Valetta I was agreeably surprised at the small amount of my hotel bill.

The mean annual temperature of Malta is 63°, the maximum 90°, and the minimum within doors is 46°. According to Rear-Admiral Smith, the average annual rainfall here is 15 inches; and the average annual number of days on which rain falls, is sixty-eight. The following table shows the temperature of Malta for one year, as observed at 9 A.M., noon, and 3 P.M. daily :—

	Maximum.	Med.	Minimum.
January, . . .	56	53½	51
February, . .	58	55½	53
March, . . .	59	57½	56
April, . . .	62	60½	59
May, . . .	71	70	69
June, . . .	75	74	73
July, . .	82	79½	77
August, . . .	82	80	78
September, . .	77	76½	76
October, . . .	70	69½	69
November, . .	65	64	63
December, . .	58	56½	55

The climate is, however, rendered variable, and therefore injurious to invalids, by the harshness and violence of the prevailing winds. These are the south-east or " Scirocco " and the north-west. The former chiefly occurs during autumn, and the latter prevails from December to March, and is a cold piercing wind laden with fine sand, and occasioning great discomfort, especially to pulmonary invalids.

It is stated that the inhabitants of Malta are generally long-lived and healthy, and Abela adduces several instances of extraordinary longevity here.

The prevailing diseases of Malta are chiefly those endemic in tropical countries, to the climate of which that of this island to some extent approximates. Fever causes a considerable mortality, and dysentery, diarrhœa, and ophthalmia are amongst the endemic complaints. Consumption is by no means as rare in Malta as might be supposed from the high temperature and southern latitude of the island. At the period Dr Hennen wrote (and the climate has in no wise changed since then), it appears that one-third of the total mortality in the hospitals of Malta was occasioned by pulmonary diseases, of which consumption caused one-fourth.

Little, however, need be said here on the therapeutic application of the Maltese climate, since comparatively few patients are now sent thither from this country. On one point, however, I have little doubt, and that is, the utter unfitness of Malta for the majority of bronchitic and consumptive patients. My reasons for this opinion are, the great tendency of this climate to sudden atmospheric vicissitudes, and the frequent occurrence of harsh and violent winds, especially from December to March, when such invalids most require a warm and equable climate.

But for the numerous class of slightly ailing valetudinarians and imaginary invalids, for some cases of chronic rheumatism, scrofula, hypochondriasis, and derangements of the digestive organs, Malta will be as useful for a change of residence and change of air as any equally warm southern climate would be, and, moreover, possesses some collateral advantages enjoyed by few other of these places.

CHAPTER XIX.

THE CLIMATES OF EGYPT.

EIGHTEEN centuries ago Egypt was to the Roman physicians and their consumptive patients what Nice and the Riviera are to ourselves at the present day. And after many ages of desuetude we find the ancient fame of the Egyptian health resorts again revived, these being now comparatively easy of access. Alexandria may be reached in six days from London *via* Brindisi; or in two days longer by the Trieste route. Invalid travellers who are not generally pressed for time would seldom be served by such rapid and fatiguing journeying, and should in preference select either the passage from Marseilles to Alexandria, or what would be still better for most consumptive patients, the voyage from England by steamer to Alexandria in fourteen or fifteen days.

The climate of Alexandria is peculiarly unsuitable to valetudinarians, and hence I shall omit any description of this city. With respect to the antiquities and monuments of Alexandria, ample accounts will be found in the ordinary guide-books; and I can only advise the invalid traveller to devote as short a time as possible to antiquarian pursuits here, and to quit this pernicious climate with as little delay as circumstances will admit. The situation of Alexandria on a low sandy peninsula between the sea and the vast swamp known as Lake Mareotis, from which there is a continual

H

malarial exhalation, is sufficient to explain the remarkable insalubrity of this town, where fevers, dysentery and ophthalmia are endemic.

The mean annual temperature of Alexandria is about 10° lower than that of Cairo. During my visit in September, the mean daily temperature was 87°, and that summer the thermometer had repeatedly risen to 115° in the shade. At night the dew-fall far exceeds that occurring in any part of Europe; so much so, that a few minutes' exposure in the evening is sufficient to wet a thick coat through. The characteristic of this climate, however, is the humidity of the atmosphere, in which respect the entire Delta contrasts remarkably with the dry climates of Upper and Middle Egypt.

"In lower Egypt," says Dr R. R. Madden, "the mummies go to pieces on exposure to external air; and in Alexandria, where the air is excessively moist, I observed several mummies melt away in a magazine where I kept them, and decomposition took place after an exposure of forty hours to the humid atmosphere; though the same bodies had resisted corruption in a dry air for perhaps forty centuries."[*]

The Rev. Dr Barclay, in his valuable work on the climate of Egypt, says—"How far Alexandria during the months of March and April may be a more suitable residence for those whose complaints require a climate at once warm, equable, and moist, I leave it to gentlemen of the medical profession to judge; but I do think myself fully warranted to denounce it as a most unsuitable place for bronchitic patients. During all the time I was there I felt as if inhaling steam; my breathing was excessively affected, and my whole system was languid and relaxed. The effects, however, by the

---

[*] "Travels in Turkey, Egypt, Arabia, and Palestine," by R. R. Madden, M.D., vol. ii. p. 76.

time I had been twenty-four hours at sea, were completely dispelled, leaving no doubt whatever as to their cause."*

My only reason for referring at all to the climate of Alexandria is that its true character does not seem to be sufficiently recognised in this country; for I have myself known instances in which consumptive patients were sent out to Egypt, and allowed to winter in Alexandria, in complete ignorance of the essential difference between the damp atmosphere of Alexandria and the perfectly dry air of Thebes. A mistake, of which the result was speedily fatal in one case, and in another proved very injurious.

Cairo is but eight hours' journey by train from Alexandria, and is built on a sandy slope between the Mokattam Hills on the east and the Nile, which, though a couple of miles distant, infiltrates the loose subsoil, rendering the western side of the city damp and unhealthy.

No eastern town that I have visted has retained its peculiar Oriental aspect so little changed as Cairo, notwithstanding its long - continued and daily - increasing intercourse with Europe. Steam and rail, which in every other land have advanced material civilization, at least, have in Egypt proved no match for the dead weight, the *vis inertiæ* of Turkish and Mahometan institutions.

The visitor, having first secured his hotel, should loose no time in ascending to the Citadel, where he will obtain a *coup d'œil* of Cairo worth all the descriptions ever penned.

Below the spectator, the outlines of the city extend over an area of more than three miles; the tapering minarets of upwards of three hundred mosques rising over the flat-roofed houses that form the narrow lanes, through which an occasional opening affords a glimpse of the bustling bazaars, crowded with men clad in every variety of Oriental and

* "The Climate of Egypt," by the Rev. Thomas Barclay, D.D., p. 63.

European garb. The peculiar clearness of the atmosphere, presenting every object with a brilliancy and telescopic distinctness unknown in Europe. Beyond the city flows the mighty river, whose sources have baffled discovery from the days of Herodotus to those of Livingstone and Cameron. On the north-east are the sites of Memphis and Heliopolis, and further still the vast outlines of the pyramids, surrounded by the desert, which extends in an interminable dreary landscape till lost in the distance.

The city is divided into quarters named after the class who reside in each. The hotels are chiefly situated in the Franks' quarter in the square of Esbekieh, and for the most part are immense half-furnished buildings. When the charm of novelty has worn off by a few days' stay, and the Citadel, bazaars, baths, and mosque of Sultan Hassan have been visited, Cairo contains nothing else to interest the tourist.

It has been stated by recent writers, that the modes of travelling are now so improved in Egypt, that every invalid who goes to Cairo may easily and safely visit the Pyramids. Being anxious, however, to prevent valetudinarians from falling into this mistake, I venture to extract from my journal an unvarnished account of a visit to the pyramids.

September 5th.—Having provided myself with a guide, and escorted by three Arabs, I started from Shepard's Hotel at nine o'clock at night, and in less than an hour's time arrived at Old Cairo, where, as the new bridge over the Nile was not then open, we aroused a sleeping ferryman who, indignant at his slumbers being interrupted by a kick in the gluteal region from the guide, refused to carry us across the river at any price. A long and angry discussion now ensued; in the midst of this altercation a guard of soldiers appeared, who, with perfect impartiality, marched us all, bystanders as well as combatants, off to the guard-house, some half-mile distant.

A small bribe having been judiciously administered by the guide, the officer on duty ordered the boatman to cross the river forthwith, and sent a soldier to enforce the injunction. We got on board, but as some fatality would have it, a new dispute now arose; the ferryman demanding ninety piastres, the usual charge being eighteenpence. Again the soldier was appealed to, and promptly settled the matter, for, entering the boat, to my horror, he commenced belabouring the man with the flat of his sword, accompanying each blow with sundry valedictory remarks on the ancient ferryman's deceased parents. This course proved very efficacious, for the price now fell to six shillings; and thus, finally, we crossed, having lost above an hour in the dispute.

As we passed through Ghizeh we encountered two marriage processions, the performers in which seemed almost mad with excitement.

Escaping from these we passed at once into the silent country, which, brilliantly lighted up by the full moon, and covered as far as the eye could reach by the inundation, now at its height, seemed to be a vast inland sea, only intersected by the narrow serpentine causeway on which we rode, and along which, every now and then, we passed by groups of fellahs lying asleep, this being the only dry ground about.

Soon after leaving the village the pyramids appeared in sight; their vast proportions, exaggerated by the uncertain moonlight, seemed close at hand. We now left the causeway, and at last entered on the dry sands of the desert, through which our donkeys struggled, plunging up to their knees. At length we arrived at the mound of rubbish which lies before the grand pyramids, weary and exhausted, at two o'clock in the morning.

Having lighted our torches, we squeezed through the narrow aperture on the north side, following the guide; and, as cautiously as possible, on hands and knees, slipped along the steep shaft, not above three feet square, which descends to the base of. the Pyramid. From this, now semi-asphyxiated, and rendered incapable of actively climbing the slippery passage, we were pushed by the guide along a gallery of marble polished to a glassy smoothness; which, after a time, expands into a lofty corridor, where I enjoyed the luxury of raising my face from the ground without risk of fracturing my cranium, though it was so steep that standing was still out of the question. I now got into the sarcophagus chamber, which assuredly did not compensate for the fatigue of the approach. The glimmering light dimly disclosed a vast apartment of granite, with a broken sarcophagus in the centre, profusely disfigured by the hieroglyphics of Messrs Brown, Jones, and Robinson, *et hoc genus omne*, some of whom have, with infinite pains, traced their illustrious patronymics with the smoke of a flambeau on the roof of the chamber.

The descent proved more facile than the entrance, and resisting the persuasive voice of the guide, who would have me visit another apartment, I struggled back again through the same aperture by which I had entered into the open air, where I arrived more dead than alive, and wrapping my cloak about me sank on the ground in a sleep of complete exhaustion. I lay thus for a couple of hours, which seemed scarcely as many minutes, until the guide warned me that it was time to be stirring. Having first visited the Sphinx, we started on our return at about half past-five in the morning, and got back to Cairo in time for a late breakfast, steadfastly resolving never again to visit the Pyramids of Ghizeh.

The mean annual temperature of Cairo is 70° and, according to the observations of Mons. Destouches, a French chemist, in

the employment of the late Pasha, the mean temperature of
the months is as follows :—

January,	. 55	May, .	. 77	September, .	76
February,	. 59	June, .	. 82	October,	. 75
March,	. 64	July, .	. 84	November, .	66
April,	. 71	August,	. 84	December, .	59

It is only during the cool season, from October to the end
of March, the mean temperature of which is 62°, that invalids
can visit Cairo, as, during the rest of the year, the climate is
so hot and arid as to be unendurable to foreigners.

Even in the depth of winter the weather in Cairo is as
mild and genial as in the latter part of spring in this country ;
the lowest temperature usually observed being in the middle
of January, when it sometimes falls to 37°. At all seasons
however, there is here a remarkable diminution of temperature after night-fall, often amounting to a difference of from
20° to 30° between the maximum of the day and the minimum
of the night. This is attributed to the fall of dew, to the
influence of the north wind, and, above all, to the extraordinary clearness of the atmosphere, favouring the radiation
of heat from the earth.

During eight months of the year cool northerly winds
prevail in Cairo. But from March to the middle of June
southerly winds, from the heated sandy plains of Central
Africa, are most frequent. One of these—the " Khamsin "—
is a perfectly dry scorching south wind, generally lasting for
fifty days continuously. During this time the lungs of those
exposed to the " Khamsin" are irritated by the impalpable
sand suspended in the air, which is then in a highly electrical
condition. The thermometer rises from 30° to 50°, and a
feeling of malaise, langour, and depression is produced
rendering any great exertion almost impossible.

With one exception, no health resort frequented by English

invalids has so dry a climate as Cairo, where the number of rainy days observed seldom exceeds twelve annually. There is, however, a heavy dew-fall at night. The atmosphere is peculiarly dry and devoid of any free humidity. The difference between Alexandria and Cairo in this respect may be inferred from the fact that the atmosphere of the latter city contains 152 times less moisture than the former.

A large number of invalids are now annually sent to winter on the Nile, most of whom suffer from chronic bronchitis or consumption. This practice might at first sight be thought a very mistaken one, when we learn that catarrhal affections, scrofula, and tubercular diseases are common in Egypt, where, moreover, phthisis occasions a large proportion of the total mortality of the native population. Whether consumption is, or is not, one of the prevailing diseases of Egypt, matters little, however, to foreign invalids, who only visit this country at the most healthy period of the year, and whose hygienic condition and general circumstances are entirely different to those of the native population, amongst whom phthisis certainly prevails.

According to the classification proposed in the introduction, Cairo must be placed in the same category with the dry, stimulating climates of Malaga, Hyères, or Nice, possessing these qualities in a much more marked degree than any of those localities.

To that large class of patients in whom incipient phthisis is insidiously progressing to its fatal issue, in young persons of phlegmatic temperament, about the age of puberty, the climate of Middle Egypt offers a resource only second in value to a long sea voyage. Cairo would also generally be a good winter resort in cases of humoral asthma, in some instances of the disease so well described by Dr Dobell under the name of " winter cough," chronic laryngeal affections, and in some cases of chronic bronchitis, and characterised by relaxation and debility. But Cairo should be particularly shunned by those who are suffering from pulmonary disease, accompanied by a hard dry cough, a tendency to congestion of the lungs, and, above all, by those disposed to hœmoptysis.

When consumption has advanced to the second stage, there are very few cases in which the climate of Cairo, or that of any part of Egypt, would be advisable ; and when the disease has reached its third and last phase, I would add, it is generally most unsuitable, although I have more than once seen consumptive patients who were almost moribund, sent out to winter, or rather to die, on the Nile.

Invalids should arrive in Cairo towards the end of October, and be prepared to leave it about the middle of March, before the unhealthy and debilitating southerly winds set in.

Upwards of thirty years ago, my father, Dr R. R. Madden, in his "Travels in the East," and Dr Richardson, in his "Travels in Nubia," published nearly at the same time, were amongst the first who pointed out the advantages of the climate of Upper Egypt for invalids, and the number of valetudinarians who have resorted thither of late years, and the testimony of recent medical writers, have confirmed the correctness of their opinions.

Unfortunately, however, Upper Egypt is only available for those who have sufficient youth and strength to endure the many inconveniences of boat life on a semi-tropical river, and amidst a semi-civilized people. The number is still further limited by the great expense which such a journey entails, so that, unfortunately, none but the wealthy can avail themselves of this, the driest and purest atmosphere in the world.

The railways from Embare to El Rodah, and from Abouxa to El Fayoum and El-Wasta, are practically useless to travellers for health in Upper Egypt, there being no proper accom-

modation for foreign invalids at any station in the valley of
the Nile beyond Cairo, so that such persons must still visit
this country in a boat, on board which they must live during
their stay in Upper Egypt.

For ordinary tourists, the Khédive Government Steamers,
which start from the new bridge " Kasr-el-Nil " at Boulac, the
port of Cairo, at regular intervals every fortnight, now afford
great facilities for an expeditious and comfortable trip as far as
the first cataract. The voyage from Cairo to Assouan and back
occupies about twenty days, the steamers stopping each night
at some village, and affording opportunities for visiting Abydos,
Thebes, Luxor, Karnack, Edfou, and nearly all the places of
interest on the river between Cairo and the first cataract.
This journey, which may be thus accomplished for £46,
although a most interesting and facile one, by no means meets
the requirements of the pulmonary invalid traveller, occupying
far too short a time, and not extending to the part of Egypt
the climate of which is most suitable for such persons, namely,
that portion of the Nile valley which lies between Assouan
and the second cataract. Hence, the pilgrim in pursuit of
health in Egypt must in most cases still follow the old plan
of river life on the Nile in a Dahabeah or large sailing-boat.
For this purpose he should arrange with a dragoman at
Alexandria or Cairo, who will undertake to provide a good
boat fitted up for passengers, with an efficient crew, and
suitable accommodation for the passengers at so much a head
per diem ; and for a party of five or six persons, who usually
club together for such a tour, the expense to each individual
would be about £50 a month.

The atmosphere of Upper Egypt, especially between
Assouan and the second cataract, is unquestionably the
driest resorted to by pulmonary invalids ; the natural
evaporation from the river being more than counteracted by

the prevailing arid winds from the adjacent desert. Professor Uhle, who passed a winter on the Nile, says,—"After ascending beyond 29° latitude the temperature rose 5° higher than it then was at Cairo. February is here the coldest and driest month; and as the spring advances, and the river falls, the atmosphere becomes still drier."

In a therapeutic point of view, Upper Egypt may be regarded as a very perfect example of a tonic, dry, warm and exciting climate, and I have already pointed out, in the first and second chapters, the cases for which such an atmosphere is adapted. Amongst these I may especially mention the early stage of phthisis and chronic bronchitis, with profuse and exhausting secretions, and occurring in persons of a torpid leuco-phlegmatic temperament, who have been brought by disease into an anæmic condition. Those disposed to congestive diseases, especially of the brain, should avoid Upper Egypt.

In every part of Egypt the invalid traveller should avoid the night air as he would poison; for there is, immediately after sunset, generally a sudden and dangerous transition, from the hot atmosphere of the day, to a chill night air, generally accompanied with a fall of dew so profuse as to supply the want of rain.

PART SECOND.

CHAPTER XX.

PRELIMINARY OBSERVATIONS ON THE SPAS.

IT would be difficult to point out a class of remedies which have so wide a range of therapeutic action on disease, and which are resorted to in so great a variety of cases, as mineral and thermal waters. The list of ailments, however, which may be benefited by the spas, would be as long and as uninteresting in this place as the "catalogue of ships" in the Iliad.

Most of those whom we meet at the watering-places are patients suffering from some form of indigestion, or from diseases occasioned by the gouty and rheumatic diathesis. These, with cases of nervous, cutaneous, and scrofulous affections, constitute the great bulk of the spa-drinkers. But besides persons labouring under actual disease, a large proportion of the patients at the watering-places suffer from no tangible malady, although obviously "out of health."

Among the throng of pilgrims to these fountains of Hygeia, come youth and age—the fagged beauty, after a season, in search of that bright complexion and those roseate hues which cosmetics cannot counterfeit; legislators and professional men; the votaries of science and of fashion; the man of business and his more laborious rival in the race of sanitary destruction, the lounger about town,—all Hadgis to the shrine of health.

In many of these cases, when a course of some foreign mineral water is enjoined, the cure which results is due not only to the active therapeutic properties of the spa prescribed, but also to the fact, that during the course nature enjoys a respite from that incessant drugging for which many valetudinarians have so unaccountable a predilection.

Another advantage of prescribing mineral waters is, that an opportunity is thus given for taking the invalid away from those habits of life which are often closely connected with his ill-health, and which, unless they are interrupted and abandoned for a time, will prevent the patient's cure. Besides this, if a distant watering-place be selected, the journey, the change of climate, of scene, and of living, often exercise a most potent therapeutic action of themselves. Lastly, it supplies an opportunity for mental rest, which is not idleness, but change of thought.

Irrespectively of these moral and indirect curative effects, most of the spas described in the following pages are, *per se*, medicinal agents of extraordinary therapeutic power, many of them being as potent for evil when misused, as for good when properly employed, and should never be resorted to without judicious medical advice.

In the succeeding chapters I have, in the first place, endeavoured to lay before my readers a succinct view of the nature and uses of mineral and thermal waters. I have next referred to the various diseases that may be treated by the springs, and pointed out the mineral water adapted to each particular case. And, finally, I have described the principal watering-places of continental Europe from my personal observations and notes taken at each spa.

CHAPTER XXI.

ON THE NATURE AND REMEDIAL EFFECTS OF MINERAL AND
THERMAL WATERS.

MINERAL springs are those which contain saline ingredients in such quantities, or in such combination, as to possess medicinal properties. They are classified into groups which bear some general resemblance in chemical composition or therapeutic action. A considerable number of the spas, however, are so complex in their character, that it would be difficult to say which ingredient so preponderates as to determine their nomenclature on chemical principles. Moreover, we find some waters containing only a trace of any chemical ingredients producing well-marked remedial effects. If, on the other hand, we base our plan of classification on the therapeutic properties of the spas, we shall be equally liable to error, since the same waters occasion widely opposite effects, according to the doses they may be taken in, the condition of the patient, and a great variety of other circumstances.

Therefore, putting aside a strictly chemical classification of mineral waters as chimerical, and passing over, for the present, the division sometimes adopted of the spas, from their medicinal effects, I shall consider the various springs described in the succeeding chapters as divisible into chalybeate, sulphurous, saline, iodated, earthy, and chemically indifferent mineral springs. Of these some are cold, and others thermal. The former contain a larger amount of gaseous and saline ingredients than the latter, and have generally a comparatively superficial origin in subterranean streams, formed by the atmospheric water absorbed by elevated mountain districts ; and are forced up to the surface by hydraulic pressure, dissolving and becoming charged with the soluble salts contained in the various strata they percolate in the transit, or with gases resulting from chemical decomposition, the chief of which are sulphuretted hydrogen and carbonic acid gases.

The effects of mineral waters are, in many instances, dependent on the amount of carbonic acid gas they contain, which not only produces physiological effects itself, but also renders substances, otherwise insoluble, capable of being dissolved. Thus, chalybeate spas, for example, are resorted to or neglected, not so much on account of the amount of iron they contain, as on the volume of carbonic acid gas by which it is rendered digestible and active.

Numerous theories have been advanced to account for the phenomena of thermal or warm medicinal springs ; and perhaps the most satisfactory of these is that which supposes their connection with volcanic action. In some places thermal springs issue in the immediate vicinity of active volcanoes. In other cases we find proof that volcanic agency once played an important part in the configuration of the neighbouring country. Such are given by the basaltic formations from which the Brunnens of the Taunus Mountains, the sources of Pfeffers, and the warm springs of the high Pyreneos issue. A still more evident proof of the relation between volcanic action and thermal springs is furnished by the fact, that direct communication seems to have been clearly established between earthquakes occurring in one country, and the hot springs of distant lands. This was more particularly the case during the great earthquake of Lisbon in 1755, when the springs of Teplitz in Bohemia,

Natres in Savoy, and Leuk in Switzerland, were simultaneously affected.

Chalybeate Waters may be divided into simple and saline ferruginous springs. The former of which, Orezza in Corsica ; Pyrmont, Driburg, Schwalbach, and Brückenau, in Germany ; Auteuil and Passy within the fortifications of Paris; and Spa in Belgium, are examples, contain little else than the carbonate of the protoxide of iron, dissolved in water more or less highly charged with carbonic acid, and are therefore not only tonic but stimulant in their effects, exciting the nervous, circulating, and digestive functions, as well as increasing the red corpuscles and fibrine of the blood.

The saline chalybeates, which generally contain some of the soluble salts of soda, together with the iron, are chiefly employed in similar cases to the saline springs when marked with debility, and in anæmia, complicated with abdominal disease, whether resulting from long residence in tropical climates, or from dietetic errors. Amongst these are the Stahlbrunnen of Homburg, Bocklet in Bavaria, Franzensbad in Bohemia, Cronthal and Rippoldsau in Germany, Saratoga in the State of New York, Tunbridge Wells, Scarborough, and Cheltenham in England.

Sulphurous Springs are those whose medicinal constitutents are sulphuretted hydrogen gas and the sulphurets of sodium or potassium. Warm sulphurous waters are invariably stimulant in their effects on the nervous, as well as on the vascular systems. The principal sources of this class are Aix-la-Chapelle, Aix-les-Bains, Baden in Switzerland, Baden near Vienna, Burtscheid, Schinznach, Teplitz, and Warmbrunn in Germany, Penticosa in Aragon, Carratraca, on the road from Malaga to Ronda, in Andalusia, and Caldas da Rainha in Portugal. The Pyrenean mineral springs nearly all belong to the same division of thermal waters.

The action of these spas is more intimately dependent on their temperature than on their chemical composition, being more stimulant the hotter they are. The warmer springs are chiefly used in gouty and rheumatic complaints, in chronic skin diseases, especially eczema, prurigo, psoriasis and syphilitic eruptions, neuralgia, and scrofulous, glandular, and arthritic swellings. The cold sulphurous springs, such as Langenbrücken, Nenndorf, and Weilbach in Germany; Bex and Schinznach in Switzerland; Castel Jaloux, Castera-Verduzan, Enghien-les-Bains and Plombièrs in France, and, nearer home, Lisdoonvarna and Harrowgate, being less stimulant, are safer, though less active in their general use, and may be employed in analogous cases to the former. Such mineral springs, however, whether hot or cold, should never be used when there is any tendency to inflammation or even congestion of any important organ, especially of the brain or lungs, and they must be most sedulously avoided in all cases in which hæmorrhage is to be apprehended.

Muriated Saline Springs are those whose active constituent is chloride of sodium or common salt. Wiesbaden, Baden-Baden, Bourbonne-les-Bains, Hammam-Melóuane in Algeria, Cannstadt, and Soden are thermal waters of this class, and Kissengen, Homburg, and the saline sources of Cheltenham and Leamington are cold. They stimulate the gastro-intestinal mucous membrane, act upon the bowels, augment the amount of urea discharged by the kidneys, increase the elimination of effete tissues, and are employed in the treatment of gout, rheumatism, scrofula, dyspepsia, habitual constipation, and torpidity of the liver.

Saline, Alkaline Waters chiefly contain sulphate and bi-carbonate of soda. Carlsbad, Marienbad, Franzensbad, and Tarasp are examples of this class of mineral springs. Their action being purgative and deobstruent, they are prescribed in

I

abdominal plethora, habitual constipation, hepatic enlargements, and gouty dyspepsia.

Muriated Alkaline sources are those in which chloride of
sodium, with carbonic acid gas and bicarbonate of soda are
the main ingredients. Ems, Salzbrunn, and Selters belong to
this category, which are all diuretic, aperient, and antacid.
Their use is indicated in the gouty diathesis and dyspepsia,
certain diseases peculiar to women, hysteria, renal and hepatic
affections.

Simple Alkaline Springs, such as Vichy and Vals in France,
Fachingen, Geilnau, and Bilin, in Germany, and Chaves and
Vidago in Portugal, contain bicarbonate of soda with excess
of carbonic acid gas. By the use of these spas the blood is
rendered more alkaline, the various excretory functions, especially diuresis, are increased, and the appetite is sharpened.
They are chiefly used in gout and rheumatism, also in gouty
dyspepsia, and in renal calculi, and gravel connected with the
uric acid diathesis.

The *Bitter Waters* are those containing a large proportion
of the sulphates of magnesia and soda. The most important
of these springs are Püllna, Saidschütz, and Sedlitz in
Bohemia; Hunyadi-Janos in Hungary; Birmenstorf in Switzerland; Friedrichshall in Saxe-Meiningen; the saline wells of
Ashby-de-la-Zouch in Leicestershire, Epsom in Surrey, and
Beulah at Sydenham, may be included in the same class of
mineral springs as the German *Bitterwässers*, acting like
them as saline cathartics, and being also diuretic and
derivative.

Earthy Springs are impregnated with sulphate, carbonate,
and chloride of lime, and carbonic acid   They are in general
thermal. Wildugen, Leuk, Lippspringe, Weissenburg, Bath,
Lucca, and Pisa, belong to this class of waters, the action of
which is astringent and stimulant. Wildugen, which con-

tains free carbonic acid, is also diuretic, and is useful in gravel and diseases of the bladder. Leuk is chiefly resorted to by patients suffering from skin diseases. Bath, Pisa, and Lucca are used in gout, rheumatism, cutaneous affections, and dyspepsia.

Jod-und-Bromhältigen Kochsalzwässer, as the Germans call springs containing iodine and bromine, generally owe their properties to iodide of sodium and bromide of manganese in a muriated saline water. Nearly all such sources are cold. The principal waters of this class are Kreuznach, Adelheids-quelle, Hall, Salzhausen, and Elmen in Germany; Wildegg in Switzerland; and Woodhall Spa in Lincolnshire; the last named being probably the strongest iodated spring in Europe. Taken internally, these waters stimulate the mucous membranes, occasion ptyalism and diuresis, promote absorption, and quicken the appetite. They are ordered in scrofulous diseases, especially of the glands and joints, in chronic glandular enlargements, such as bronchocele, in secondary syphilis, and, above all, in the treatment of some of the diseases peculiar to women, connected with chronic uterine or ovarian inflammation.

The so-called *chemically indifferent springs*, such as Schlan-genbad, Teplitz, Gastein, Neuhaus, Tüffer, and Wildbad in Germany, Chaudefontaine in Belgium, and Pfeffers in Switzer-land, are thermal sources, some of which contain absolutely less mineral matter than our ordinary drinking water. Thus, the New River water contains 2½ grs. of solid matter to the pint; the East London Company 3 grs.; and that supplied by the Hampstead Company 4½ grs. Yet these waters cause no apparent effect, whilst the springs of Wildbad, with 3½ grs. of salt to the pint; Pfeffers and Gastein, 2 grs.; and Chaude fontaine, with 2¼ grs., are all capable of producing therapeutic results, which seem to be mainly due to the temperature at which they are employed. These chemically indifferent baths, especially the cooler ones, possess peculiarly sedative effects, not only allaying nervous irritation, but also diminishing vascular excitement, whilst the warmer springs of the same class, being more stimulant, are chiefly used in aggravated cases of chronic rheumatism and rheumatic-arthritis.

We may now proceed to consider the therapeutic application of these various mineral waters, which is by no means solely determined by their chemical composition, as is supposed by some able chemists ; one of whom goes so far as to assert that it is only necessary to ascertain what are the saline ingredients in any mineral water, "and dissolve the same con-stituents in the proper proportions, and you regenerate the spa to all intents and purposes." I presume, however, that the "intent and purpose " of the invalid who resorts to any spa is the cure of some ailment; and most assuredly this will not be accomplished by the "simple method " of the chemist. The baths of Wildbad, for instance, contain only two grains of common salt, with half a grain each of carbonate and sulphate of soda to the pint of water, at the temperature of 98° ; and yet wonderful cures of chronic arthritic diseases are effected by a course of them. But very few practical physicians could be persuaded that an artificial solution of the same salts would produce the same effects. If it had the same influence, would it not be as absurd as cruel to send a patient, crippled by chronic rheumatism, to a remote village in the Black Forest, when a handful of common salt, with a teaspoonful of washing-soda and Epsom salts, would convert a tub of luke-warm water in his own bedroom into a *verjüngen*, or " youth-restoring fountain ? "

Without entering into the much disputed question of the mode of action of chemically indifferent and other mineral waters, concerning which many German writers seem to revel in explaining the *ignotum per ignotius,* each successive 'Brunnenartz' putting forth some conjecture if possible more misty and unintelligible than that of his predecessor, I believe that we may content ourselves with the empirical knowledge that certain mineral waters produce certain effects. In many instances these waters act simply as so much pure water would do could it be swallowed under the same circumstances, and with the same anticipation of success which mineral waters are used with. When we consider the amount of water consumed daily by each patient at most spas, we will find that as diluents these waters must have a considerable action on the animal economy. Ten or twelve pints of mineral water per diem is a common dose at several of the spas, and it is obvious that the mere passage of so large a quantity of fluid through the system must break down and wash away morbid deposits and disordered secretions. The effects of the journey to the spas are oftentimes sufficient to account for the benefit ascribed to the mineral springs. The mode of life at most continental spas is usually totally different from the ordinary habits of the British valetudinarian, who is forced into earlier and more regular hours, and into taking more exercise than he has been accustomed to; and this change is probably one of the most important advantages of the foreign watering-places.

The majority of those who resort to the continental watering-places suffer from disorders connected with the gouty diathesis, or with some gastric derangement. Now, travel as we may, whether we " rough it " on foot through the mountain bye-roads of Switzerland or Spain, or sedulously avoid fatigue in the sleeping-car of a train, more or less exercise must be taken; and oftentimes it is this exercise which does the good ascribed to the mineral water or change of climate. By it the circulation is quickened, but at the same time is equalised ;

that is, all the vessels receive more blood, and thus the amount accumulated in any one organ is diminished. The respiration is also hurried, and therefore more carbon is exhaled from the system. The appetite is sharpened. The change of diet, the freer use of fruit, the light acidulous wines, the oleaginous cookery, all promote the alvine and renal excretions. The "moral" influence of the change, too, conduces to the physical improvement it effects. New scenes and places suggest new thoughts; the *atrabilis* of gloomy apprehension is purged away, and the patient, ceasing to think on his symptoms, they cease to exist.

Still I am very far from asserting that change of living and scene will of itself produce all the benefits often effected by a journey to, and use of, an appropriate spa. No amount of travel will, unaided by other treatment, purify the vitiated blood of a gouty patient, reduce to proper proportions a tumefied liver, unbend a contracted joint, or impart vigour to a palsied limb, all of which cures are oftentimes performed at the spas.

Though most of the saline constituents of mineral waters are articles of the Pharmacopœia, and generally exist in comparatively small quantities in the spas, yet they act more effectually, owing to a more perfect solution and finer division of the natural preparation than can be effected by artificial means. Another reason of this superiority is, that spas occasionally contain substances insoluble in ordinary water. The Homburg springs, for instance, contain carbonates of lime, iron, and magnesia, and silica, which are rendered soluble by the presence of free carbonic acid gas.

The mode of using, and dose of the spas, depends on the composition of the particular spring selected, the condition of the patient, and a variety of circumstances, which render it impossible to lay down any accurate general rules on these points. Still I shall attempt a few suggestions which may be found useful.

In the first place, mineral waters should, if possible, be drank at their source. The patient who uses the imported mineral waters sold in this country, not only loses the chief benefit of the prescription, that is, the change of scene, of climate, and of living, but further, the mineral fluid is in most cases changed and deteriorated by keeping.

The spa physicians, as a rule, advise their patients to disregard all other medicines, and confide in the virtues of the mineral water of their place of abode. It would be rather difficult to say whether this recommendation has done most harm or good. In many cases it is, doubtless, a wholesome counsel. But, on the other hand, when the patient is really ill, the use of mineral waters should not supersede all other treatment; for, as a very observant English physician of the last century well remarked, " it is but prudent to bring all the forces one can raise against so formidable and so potent an enemy as a chronical disorder."*

The quantity of water to be consumed is a matter that should be prescribed to the patient before he is sent to the spa, and must be determined by the circumstances of each case. Invalids are generally left to their own discretion on that point, and seldom exercise any. Not a few valetudinarians think that the benefit to be derived is in exact proportion to the amount of mineral water they can swallow. As a rule, two glasses of water before breakfast, one before dinner, and a couple of tumblers-full in the evening are quite sufficient.

In cases of structural cardiac, pulmonary, or cerebral disease, mineral, and especially thermal waters are generally

---

* Cheyne, " An Essay on the Gout and the Bath Waters," p. 59, London, 1721.

worse than useless. Consumptive patients are constantly sent to certain spas, and we are assured that cures are thus effected. My own belief, founded on the study of this disease in many climates is, that no case of phthisis was ever cured by any mineral water, although the general health of phthisical patients may probably sometimes be benefited by a course of light chalybeate or alterative waters.

Natural warm or thermo-mineral baths may be divided into two groups, the first of which have a temperature ranging from 85° to 97,° or lower than that of the blood; and the second comprising all the baths whose heat exceeds this. The effects produced by baths of the former kind are of a sedative character; after a few minutes immersion the nervous system is soothed, pain and irritation are allayed, a feeling of physical comfort and tranquillity is induced; the functions of the skin, however, become more active, the various secretions are increased, and changes are produced in the blood, by the transudation and absorption that take place in the bath, which are greatly influenced by the chemical composition and density of the water.

The principal maladies treated by baths of this class are—chronic nervous and spasmodic affections, such as neuralgia and sciatica, subacute disorders of the abdominal and pelvic viscera, especially of the liver and gastro-intestinal mucous membrane. This remedy is especially applicable to the treatment of many of the diseases of women connected with chronic inflammation of the uterus or ovaries, in some cases of painful or defective menstruation; and in hysteria dependant on these causes. Thermal baths are also largely and successfully employed in rheumatic affections; and, finally, these baths are used with remarkable success in the treatment of· chronic skin diseases.

The second class of thermal baths comprehends all those whose temperature is higher than that of the blood, and ranges from 98° to 120°. Such baths are strongly stimulant, their exciting power being in proportion to the elevation of their temperature. The blood is determined to the surface by them, the cutaneous capillary vessels become congested, the respiration is hurried, the pulse is accelerated and becomes full and throbbing, and a sense of oppression and discomfort is felt until a profuse sweat relieves the circulation. When this has broken out, all the uneasy sensations gradually disappear, and are succeeded by a sense of languor and exhaustion, with a tendency to sleep.

Widely spread as they are throughout Europe, the therapeutic uses of hot mineral baths is limited to a small class of patients. They are employed in certain obstinate cutaneous affections, in chronic rheumatism, and rheumatic arthritis.

The length of time the patient should remain in the bath depends on the nature of the mineral water, and on that of his disease, and varies from ten minutes, which is the usual period of immersion at some of the hotter sulphurous baths, to eight or ten hours, which formerly at Pfeffers was thought an ordinary occurrence.

The ordinary duration of a bath at Wildbad, Aix-les-Bains, Gastein, and several other places, is an hour. So prolonged an immersion in a fluid at the same temperature as the blood stimulates the cutaneous capillaries, withdraws the blood from internal congested organs, allows their vessels to recover their healthy tone and contractile power, and must be capable of producing powerful curative results.

Care should be taken to avoid exposure to cold immediately after leaving the warm bath, and therefore I think that the very early hour at which the bath is generally taken at the spas of Germany is open to some objections; and in many cases it will be found better to bathe in the middle of the day three or four hours after breakfast, or in the evening, about five or six hours after the early German dinner.

Whenever a patient exhibits any apoplectic tendency, hot mineral baths are most dangerous, and might lead to sudden and fatal results. Very fat people should be equally cautious of indulging in such baths. In all cases, too, in which the heart, aorta, or any important vessel is diseased; and, indeed, whenever it is necessary to guard against vascular excitement, thermal mineral waters are contra-indicated.

CHAPTER XXII.

DYSPEPSIA AND THE SPAS.

WITH the exception of the maladies produced in this climate by wet and cold, most of the diseases for which health resorts are visited are the result of an excessive or injudicious diet,— "plures crapula quam gladius." The object of the present chapter is to show that most of these complaints, including all cases of dyspepsia whose leading symptoms are impaired, or fastidious appetite, painful, slow, or imperfect digestion, irregular action of the intestinal canal, and hypochondriasis, may be better treated by mineral waters, conjoined to abstinence, or some alteration of living, than by pharmaceutical preparations.

On no subject whatever have more numerous and contradictory systems been propounded than on diet, and on none would it be more utterly impossible to lay down any universally useful general rules than on this. The majority of people eat far more than nature requires, and far more than they can assimilate ; and, consequently, very few enjoy perfect health. But the number of meals and the quantity of food consumed can only be decided by each individual's experience of what agrees with himself; for some would gain flesh on Carnard's allowance of eleven ounces of food per diem, whilst others would starve on as many pounds. Therefore the only rule I would venture to recommend to the dyspeptic patient is that, putting aside the bugbear of debility, he should use only the quantity of food he can digest, without pain or discomfort, however small this may be.

Although we have progressed vastly from the bibulous habits of our grandfathers, still, to the present day, far too much wine is used by the upper classes in this country, and many of the ailments by which life is embittered, and premature death induced, and to ward off which the spas are resorted to, are owing to this cause. Wine and all other alcoholic liquors are quite unnecessary to a man in perfect health; but comparatively very few are to be found in that happy state. If we lived in a world in which no sorrows, cares, or anxieties of the mind, reacting on its frail tenement the body, produced diseases, and paved the way to death, then would I join those who preach total abstinence from all intoxicating drinks. But as, unfortunately, we live on earth, and not in Utopia, and have to fight the battle of life often against unequal odds, with failing strength and sinking heart, therefore, with Solomon, would I prescribe—" Wine to him that is ready to perish, and to him that hath grief of heart." Two or three glasses of wine daily, however, are quite sufficient for any adult, and under no circumstances whatever should wine be given to a child, except as a medicine.

To treat dyspepsia, we must bear in mind its causes; and these, besides errors of diet, are severe mental labour, conjoined to a sedentary life, nervous excitement and anxiety, the abuse of tobacco or tea, and late and irregular hours. These must all be corrected, and a journey to, and residence in a distant watering-place, will often prove the most efficacious mode of effecting this. The travelling, with its concomitant change of scene and of climate, benefits the patient's general health, takes his attention from his ailments, and thus proves an antidote to the gloomy and depressant influence of dyspepsia. The early and regular hours observed at the continental spas are a wholesome contrast to the late hours of fashionable life, where the example of Smyndiris, the Sybarite, who for twenty years never saw the sun rise or set, is very frequently followed. Lastly, the mineral waters themselves are remedies more powerful than any prepared by the apothecary in the treatment of indigestion and hypochondriasis.

Most of the spas resorted to by dyspeptic invalids are saline springs, of which Soden, Homburg, Wiesbaden, and Kissingen are perhaps the most generally employed. Occasionally the more stimulating and cathartic springs, such as Franzensbad, Carlsbad, and Marienbad, exert a happy influence on derangements of the liver connected with indigestion. Often, when this is obviously dependent on nervous debility and weakness, the chalybeate saline waters are indicated; but so protean are the forms of indigestion, that it would be impossible here to enlarge on this topic. Whatever spa may be chosen, the patient should be taught to depend more on his own self-control than on the mineral water for his cure.

CHAPTER XXIII.

ON GOUT AND ITS TREATMENT BY MINERAL WATERS.

ALTHOUGH an elaborate disquisition on the gout would be misplaced in a work of this kind, still a few preliminary observations on the disease appear to me necessary for understanding the action of certain spas in its cure.

Formerly gout was deemed a sign of wealth, and even of a vigorous intellect. Thus Sydenham remarks that "gout destroys more rich than poor, and more wise men than fools." Whatever may have been the case then, at present, however, gout attacks rich and poor with impartiality; and were it confined to the wise alone, the fees of gouty patients would, I fear, be considerably diminished. The fact is, that whenever people of any class indulge largely in animal food and fermented liquors, gout will prevail; though, of course, the injury caused by the excessive use of azotized articles of diet will be less in proportion to the amount of exercise or labour performed. But though the gout may be produced or engendered in almost any individual by certain habits of living, more frequently it is the result of hereditary transmission. A tendency to gout or the gouty diathesis, descends for generations; and the inheritor of these tendencies, unless he use precautions very few submit to, when he attains middle age, will experience the effects of his own and his ancestors' indiscretions in a fit of the gout.

An attack of regular gout is generally preceded by gastric derangement, diminished excretions, languor, groundless depression of spirits and great irritability of temper. When such symptoms occur in a man of full habit of body, beyond the middle age, and especially if he be of gouty parentage, we may safely conclude that podagra will declare itself in a few days. If nothing be done to stave off the approaching seizure, after three or four days' slight indisposition the patient may go to bed in his usual condition, and awake from a sound sleep to suffer the torture of acute gout. For some nights the pain and febrile disturbance recur, though with less severity, the cuticle peels off, the urine becomes more copious, depositing lithic salts, and the attack gradually wears away. The interval which now elapses before the next seizure after a first fit of gout will probably last a year, but becomes shorter after each recurrence, until at last the invalid is hardly ever free from the disease.

Besides what is known as the regular gout, the disease in question plays a protean part in the chronic ailments of the richer classes in this country. We meet every day patients who have never complained of gout, but who in point of fact are never free from that disease. These persons are, for the most part, men of enfeebled constitution, of a sallow, cachectic complexion, with a rough, dry skin, an irregular, often feeble and compressible pulse, scanty excretions, variable appetite, complaining of frequent heartburn and uneasiness in the right hypochondrium, of irritable disposition, and despondent temperament. In this gouty habit of body, mineral waters offer the most appropriate remedy.

In every age satirists have found a subject for ridicule in the multiplicity and uselessness of the remedies employed for the treatment of gout. Ovid emphatically asserts that—

" Tollere nodosam nescit medicina podagram."

Nor was Fenton, who died himself from this disease, more complimentary to the power of physic when, in his " Ode to the Gout," he addresses it as—

> " Thou that dost Æsculapius deride,
> And o'er his gally-pots in triumph ride. "

Sydenham very tersely sums up all the remedies used for gout in his day, and, though a hundred and eighty years have elapsed since, his conclusion is still to some extent apposite. He says—" In gout, too, but three methods have been proposed for the ejection of the *causa continens*—bleeding, purging, sweating. Now none of these succeed." But during an acute paroxysm of gout, it is seldom easy for the sufferer to content himself with flannel and patience ; and, fortunately, a great deal may be done in most cases in the way of mitigating the pain, and expediting the recovery. Herein lies the skill of the physician. The charlatan treats all cases alike, and, as an old writer well expresses it,—" Sometimes he kills the disease ; but more frequently he kills the patient."

We now come to the treatment of gout by mineral waters. It is obvious, however, that this treatment can only be used as a prophylactic, or in the intervals between the paroxysms. By the proper use of mineral waters we open avenues through the excretory organs for the elimination of those principles whose presence in the blood gives rise to the phenomena of gout. The secretions from the intestinal mucous membrane, kidneys, and skin, are all augmented ; and through these channels the gout-producing lithates are washed away. When the constitution is weakened, by repeated seizures of gout, chalybeates are often required ; and in such cases the superiority of nature's pharmacy to that of art is shown ; for oftentimes, when every preparation of steel has been tried without benefit, some of the mild chalybeate springs "work wonders," restoring tone and strength; and thus it is that Schwalbach, Spa, or Tunbridge Wells prove useful, by giving strength to the system, to localise and develop regular podagra, in the stead of that misplaced, atonic, irregular, wandering form of gout now so prevalent.

The carbonated alkaline springs, such as Vichy, Bilin, Fachingen, and Geilnau, are most commonly prescribed in the intervals of regular gout. They correct the acidity of gouty blood, rendering it and the urine alkaline, facilitate digestion, and increase the secretions. The simple saline or muriated saline springs are also largely employed in gouty cases, and especially in gouty dyspepsia. They stimulate the appetite and digestion, and thus improve the blood. Their effects vary according to their temperature, some of them being cold and others thermal. Homberg, Cheltenham, and Kissingen belong to the former, and Wiesbaden, Baden-Baden, Soden, and Cannstadt are examples of the latter. The saline alkaline spas, such as Carlsbad, Marienbad, and Franzensbad, are occasionally useful in cases of irregular gouty disease in persons of full plethoric habit.

Of the English spas resorted to by gouty sufferers, perhaps the most generally applicable are the warm saline waters of Bath; next rank those of Buxton; then the chalybeate-salines of Cheltenham and Tunbridge Wells.

Whatever may be the mineral water selected in the treatment of any case of gout, it should be borne in mind that its curative influence must be aided by the same abstemious and guarded mode of living which the patient would have required had he been treated at home.

CHAPTER XXIV.

ON MINERAL WATERS IN THE TREATMENT OF SOME OF THE
DISEASES PECULIAR TO WOMEN.

THE importance of constitutional remedies, and the efficacy
of certain mineral waters in the cure of most of the diseases
peculiar to women, are subjects which I have elsewhere
discussed, and I now venture to reiterate very briefly my
views on these points.

Within a comparatively recent period the diseases and
displacements of the uterus have come to occupy the most
prominent place in the etiology of female diseases, and it
may be fairly asked, whether these complaints have really
become more common than was formerly the case? Or, is it
merely the fashion of the present day to ascribe all obscure
female ailments to uterine causes? Or, have these been
always as prevalent, though only now discovered by the
improved means of diagnosis furnished by modern gynæco-
logical science?

Uterine complaints are certainly more in vogue as well as
more easily diagnosed than was the case a few years ago.
But irrespective of this, however, they must also be admitted
to be more prevalent; owing, as I believe, mainly to the
increasing luxury of the present age, and the artificial habits
and conditions of modern life, producing gout, and, indirectly,
scrofula—the most frequent, although generally unrecognised,
causes of the diseases peculiar to females.

Uterine specialists may be considered as followers of one or other of three distinct schools of gynæcology. A few years ago most of the diseases peculiar to women were commonly attributed to inflammation of the cervix uteri; then newer views as to the frequency and pathological importance of ovarian disease in such cases prevailed; and, lastly, the doctrine that these complaints are the result of displacements and flexions of the uterus has come to be widely adopted as the only sure foundation of uterine therapeutics;

> " For in physic, as in fashion, we find
> The newest has always the run of mankind."

Each of these theories is undoubtedly applicable in many instances, but neither of them holds that exclusive and primary place in the causation of the diseases peculiar to women which some writers insist on.

From my own experience, I would say that the most frequent immediate cause of impaired female health is chronic uterine, or utero-ovarian inflammation. Thus, rather more than one-tenth of all the patients under my care at the Dispensary for Diseases of Women suffered from endo-metritis or cervicitis, and in private practice I have found the proportion of these cases fully as large. The consequences of chronic inflammation of the womb and its appendages are as important as its frequency. In some instances chronic metritis occasions hypertrophy and ulceration of the cervix and os uteri, vaginitis and leucorrhœa; in others congestion and enlargement of the fundus, eventually causing flexions and displacements of the womb; and in others again it extends to the ovaries, producing menstrual irregularities, sterility, and hysteria in all its forms.

Sterility almost always accompanies this disease, and as long as it exists to any serious extent the patient must remain barren. This fact, which I regard as one of great importance, is too generally ignored in practice. I have known instances in which patients were subjected to very heroic surgical treatment to overcome some supposed mechanical obstacle to impregnation, and who nevertheless remained childless, no attention having been paid to the true and most frequent cause of sterility, namely, the existence of chronic inflammation, on the subsequent cure of which pregnancy has immediately followed.

Ovarian inflammation, manifested by soreness, tumefaction, and occasionally burning pain in the ovarian region, is one of the most common consequences and accompaniments of endometritis. In these cases the inflammation extends from the uterus, along the Fallopian tubes to the ovaries; and hence patients thus affected are sterile for the time being.

The treatment of the affections now under consideration is still vague and unsatisfactory, generally extending over long periods of time and often unrewarded by the-cure of these diseases, their predisposing causes being, as I believe, overlooked in practice.

Of the predisposing causes of chronic inflammation of the uterus, by far the most frequent is the scrofulous diathesis. Some years ago I observed, and called attention to the fact, that a large proportion of the patients attending the Dispensary for Diseases of Women were of a well-marked strumous habit of body, or actually suffered from glandular or cutaneous scrofulous affections. In such cases uterine complaints are necessarily impressed with the constitutional taint.

Women are supposed to be in a great measure exempt from gout. This opinion is certainly unfounded with regard to anomalous gout, which attacks women quite as much as men; and, in the former, affects the uterus as commonly as regular gout does the joints in the latter. There is no reason why the gouty diathesis should be confined to men. Gouty parents generate female as well as male children; any hereditary disposition must be shared in by both alike, and the exciting causes of the disease are obviously not limited to either sex.

Many of the symptoms which Gooch described under the name of irritable uterus, and which modern gynæcologists have transferred to the account of displacements of that organ, are oftentimes produced by gout, and may occur independently of any mechanical cause, though they may also coexist with it. Thus, in a paper on this subject, which I published some time ago in the *Lancet*, I instanced the case of a widow lady, aged about thirty, who had for years been suffering from lumbar and pelvic pain, extending down the left thigh, and accompanied with lameness, dysuria, and slight uterine catarrh. The cervix uteri was congested and the os patulous, the sound penetrated five inches, and the fundus was tilted backwards. Before she consulted me she had been treated by others for retroversion by mechanical expedients only, and for a long time I did the same. Every pessary,—and the number was almost countless that I tried,—however well it might fit, was practically useless: before a week's time she would limp into my study in as bad a plight as ever. I need not go through the details of the case further than to add, that I learned at last that she inherited gout, and, on examining, found her urine laden with uric acid. She was then treated by alkaline remedies and colchicum, and sent to Vichy, whence she returned with all the symptoms relieved, and, though yet necessitated to wear a Hodge's pessary, is practically free from any discomfort.

With regard to active local treatment in ordinary cases of chronic inflammation and simple ulceration of the os and cervix uteri, Talleyrand's advice might be advantageously adopted by gynæcologists—" Surtout point de zele." If we trusted more to constitutional remedies, and above all to the judicious employment of certain mineral waters, in such cases, I verily believe that in many instances our patients would get well sooner than they do, when the local irritation is increased *secundum artem* by frequent examinations and the repeated application of escharotics.

In considering the uses of mineral waters in cases of impaired health connected with chronic inflammation of the uterus and its appendages, we should in the first place recognise the fact that local treatment, although not to be neglected, should be subordinate to the cure of the constitutional disease which is the remote, and too commonly the undetected, cause of the local complaint. Thus, in scrofulous or gouty uterine disease, the faulty state of the blood should be corrected by alteratives and tonics, or by antacids, and above all by the natural chalybeate and iodated mineral waters, or by the alkaline carbonated spas, as the case may be.

Hysteria in some form is generally associated with chronic uterine disease, and this underlies and complicates most of the symptoms for which gynæcologists are consulted. Counterfeiting every malady, acting through and upon the nervous system, attended with groundless apprehension, depression of spirits, and morbid irritability of temper, oftentimes rendering the patient herself as miserable as she renders those about her, this disease is closely allied to that graver nervous lesion which constitutes insanity, and, if unchecked, may pass into it. Local treatment, except to rectify some displacement or subdue well-marked ovarian or uterine inflammation, is of little utility in such cases; nor are the tonics and antispasmodics usually relied on, comparable in their therapeutical effects in restoring a hysterical woman to the *Mens sana in corpore sano*, to the saline chalybeate waters, such as

Ems, Schwalbach, or Spa, provided these be used at their source.

The use of a remote spa in these cases is something beyond the benefit to be derived from the mineral water. The functions of the liver and bowels, commonly torpid in hysterical women, are stimulated by the change of living, and a sedative effect is generally produced on the hyperæsthetic condition of the patient. The changes and incidents of the journey suggest new ideas, by which the patient's mind is diverted from that morbid concentration on her ailment which characterises hysteric disease.

Amongst the mineral waters that may be employed in the treatment of chronic uterine complaints, the iodated and bromated saline springs, such as Wildegg, Woodhall-Spa, Kreuznach, Adelheidsquelle, Hall, and Salzhausen, deservedly hold the foremost place. These waters act as special stimulants to the mucous membranes and glandular system, promote absorption, occasion ptyalism and diuresis, quicken the appetite, and produce the resolution of glandular enlargements. Hence their singular efficacy in the treatment of the diseases of women produced by chronic uterine enlargements and hypertrophy, the result of congestion or chronic inflammation of the womb; and especially in these cases of sterility which are supposed to be occasioned by hypertrophy of the cervix uteri.

The second class of mineral waters applicable to the treatment of the chronic uterine diseases now under consideration are the chalybeates, both simple and saline. The former are those most resorted to by sufferers from chronic diseases of the womb, and are especially adapted for the treatment of chronic ulceration of the cervix uteri and uterine or vaginal leucorrhœa, associated with anæmia, as well as in the constitutional debility and loss of tone so frequently produced by,

152   MINERAL WATERS IN THE TREATMENT OF

as well as conducive of, uterine irritation, [inflammation, congestion, or ulceration. Chalybeate spas also exercise a marked curative action in cases of hysteria dependent on these causes, as well as in certain instances of sterility. The principal simple chalybeate waters suitable for such cases on the Continent are Spa, Pyrmount, Brüchenau, Schwalbach, and Driburg.

The saline chalybeate springs may also be used in various forms of chronic uterine disease producing anæmia and complicated with abdominal and other enlargements, and, according to my experience, are particularly serviceable in the uterine disorders so commonly caused in European women by tropical climates, and especially by long residence in India. These springs generally contain the salts of soda in combination with iron, and amongst them those most suitable for the cases we are now considering are the Stahlbrunnen of Homburg, Franzensbad, Bocklet, and at home, Tunbridge Wells and Cheltenham.

Sulphurous mineral waters are the third class which I regard as applicable for the treatment of the uterine diseases above referred to. Thermal sulphurous spas being strongly stimulating, can only be used in cases where the patient's constitutional state is not plethoric, and where there is no danger of enkindling latent inflammation, and thus converting a chronic into an acute disease. The warm sulphurous springs that are available for the treatment of chronic inflammation of the womb are Schinznach in Switzerland, Baden on the Limmat, Aix-les-Bains, Eaux-Bonnes, and Amelie-les-Bains. Cold sulphurous waters, such as Lisdoonvarna, Harrowgate, Enghein-les-Bains, and Plombières, may also and with greater safety be employed in some cases of chronic uterine inflammation.

Whenever uterine and ovarian dysmenorrhœa, pain, or

any other evidence of inflammation is present, · there are no remedies of such universal applicability as the chemically indifferent thermal baths, such as those of Pfeffers, Schlangenbad, Gastein, Wildbad, and Chaudfontaine, all of which exercise a powerful sedative effect on the nervous and vascular systems, and are especially suitable for cases of uterine disease associated with hyperæsthesia and hysteria, or abnormal nervous susceptibility. Besides these, the thermal arseniated waters of Mont Dore and St Nectaire, both in the volcanic district of Auvergne, may be used in uterine disorders of scrofulous or neuralgic origin. The warm mineral waters of St Sauveur, in the Eastern Pyrenees, which, in addition to their high temperature, contain a large amount of the peculiar pseudo-organic unctuous substance termed " glairine " or " baregine," have a great and, I believe, well-merited reputation in France in the treatment of scrofulous, rheumatic, and neuralgic affections, as well as in hysteria, leucorrhœa, and other complaints peculiar to women, resulting from chronic uterine disease.

It can hardly be necessary for me to observe that, although I attach so much importance to the constitutional treatment of uterine maladies, which, I believe, is too generally overlooked at the present day, I am by no means insensible of the equal importance of conjoining efficient local treatment with the constitutional remedies indicated in such cases.

CHAPTER XXV.

CHAUDFONTAINE.

THE first Belgian spa on the route from England to the Rhine
is Chaudfontaine, and if facility of access, beauty of situation,
and convenience of living were the chief recommendations for
a watering-place, then should Chaudfontaine be among the
most frequented spas of Europe.

Chaudfontaine may be easily reached in fourteen hours from
London; but I should myself be very loathe to travel with
the man who on any but the most urgent necessity would
thus rush through Belgium, of which the majority of health
travellers, armed with through tickets from Victoria or
Charing Cross to Homberg, or whatever German spa they
intend to visit, now generally see little or nothing. For my
own part, I love to loiter along the ancient streets of Bruges
or Ghent, or, losing my way in the quaint winding thorough-
fares of Antwerp, to stray through its glorious cathedral and
churches. Therefore I confess that I cannot relish being shot
by express train from country to country in the cannon-ball
fashion of travelling now so much in vogue, and more
particularly through a country like Belgium, where every
town—nay, every village—presents remains of historic interest
that can be found in no other part in Europe.

Within a quarter of an hour's journey of Liège, in the
valley of the Vesdre, surrounded by beautifully wooded hills, is

the village of Chaudfontaine, consisting of a long straggling street of small hotels and lodging-houses. Immediately opposite to the railway station is the thermal spring, bath-house, and assembly rooms. This source belongs to the class of chemically indifferent springs, owing its effects chiefly to its temperature, which is 92°; the saline ingredients contained in a pint of the water being but a grain and a half of common salt, and rather less carbonate of lime.

The physiological effect of the Chaudfontaine baths is sedative. The duration of these baths should not exceed twenty minutes, and they are principally employed in cases of neuralgia, chronic rheumatism, and contractions of the joints from this cause. I can also recommend their use in some of the diseases peculiar to women depending on uterine or ovarian hyperæsthesia, congestion or chronic inflammation, and more especially in some forms of painful and imperfect menstruation. The internal use of this water is also resorted to as an adjuvant to the baths, which are contra-indicated in all cases in which any tendency to hæmorrhage exists.

One great defect in these bath-rooms is their imperfect ventilation, so that they become filled with steam which, con-densing on the walls, runs down in a stream on the bather's garments. The season commences in May, but the most crowded month in Chaudfontaine is July.

CHAPTER XXVI.

SPA.

OF all the foreign chalybeate mineral waters resorted to by English valetudinarians, the most accessible as well as some of the most efficacious are those of Spa. This watering- · place, which is separated from London by only sixteen hours of easy journey, is charmingly situated in a valley in the Ardennes Hills, by which it is sheltered from northerly and easterly winds. The town itself is merely an aggregation of hotels and lodging-houses, together with a few shops, as the resident population is barely four thousand people, who contrive to live by supplying the wants of the thirty thousand visitors who annually drink these waters.

Spa is one of the most enjoyable watering-places on the Continent. The beauty of the surrounding scenery and the many easy and picturesque walks and drives in the imme-diate vicinity render this place peculiarly adapted for hypochondriacal and dyspeptic valetudinarians; whilst those who do not consider open air exercise in a fine climate and amidst beautiful scenery a sufficient enjoyment, will find in Spa a theatre open on Sundays, Tuesdays, and Thursdays, a public ball on Wednesdays and Saturdays, and a concert on Mondays and Fridays, during the season.

The living at the hotels—of which there are some twenty-four—is better than at most of the Continental watering-places. The small mutton of the Ardennes is equal to the Welsh, and the neighbouring streams furnish an abundant supply of excellent trout, carp, and roach.

The season lasts from the first of May until the end of September, when the weather generally becomes cold and broken.

The Spa waters are powerful chalybeates containing, in addition to the carbonate of the protoxide of iron, a large volume of free carbonic acid gas. Hence their action, not only tonic but stimulant, and their use in the treatment of almost all chronic diseases of an anæmic character in which chalybeates are indicated.

Analysis of the principal Mineral Springs of Spa.

Sources.	Temperature.	Carbonic Acid Gas in Cubic Inches.	Solid Ingredients in Grains.	Carbonate of the Protoxide of Iron.	Carbonate of Lime.	Chloride of Sodium.	Carbonate of Soda.	Carbonate of Magnesia.	Silica.
Pouhon,	50	21	3·37	0·87	0·75	0·20	0·90	0·31	0·28
Geronstère,	49	14	1·65	0·45	0·33	0·09	0·45	0·16	0·10
Sauvenière,	49	20	1·70	0·43	0·22	0·06	0·30	0·10	0·07
1st Tonnelet,	49	22	0·96	0·39	0·15	0·04	0·21	0·08	0·04
2d Tonnelet,	...	19	0·58	0·25	0·12	0·01	0·08	0·06	0·02
Groesbeck,	...	21	0·83	0·24	0·16	0·04	0.22	0·08	0·04

The Pouhon, which is the mineral source *par excellence* of Spa, as it contains twice as much saline ingredients as any of the other springs, and is the only one exported, is situated in the centre of the town. It rises through a handsome marble basin, around which each morning a crowd of valetudinarians congregate, struggling to reach the presiding Naiades, two ancient females who deal out the water in exchange for a few sous. The spring issues from a soft ferruginous slaty rock; is clear and sparkling, and though chalybeate is decidedly piquant and agreeable in taste.

Besides cases of simple amenorrhœa connected with anæmia

and chlorosis, the Pouhon may be used by patients whose convalescence after childbirth has been slow and imperfect.
This spring is especially recommended by Dr Sutro "in
obstructed portal circulation, in deficient bilification, in congested liver and spleen, following intermittent fever; also in
flatulency, digestive weakness and acidity, tendency to
diarrhœa and passive hæmorrhage."* It is also occasionally
employed in cases of dropsy supervening on acute diseases,
and it was in this malady that Peter the Great, a pompous
inscription to whose memory adorns the "Temple de Pouhon,"
found the Spa waters so efficacious.

A mile and a half distant from the town is the source *de
Barisart*. This spring, though weaker than the Pouhon in
saline constituents, contains a larger volume of carbonic acid
gas, and may be prescribed with greater advantage in some
instances of extreme debility after recovery from fever, and in
certain cases of hepatic obstruction and also in uterine catarrh,
resulting from chronic endo-metritis of scrofulous origin.

One of the most interesting sources of this watering-place,
at least to gynæcologists, is the *Sauvenière*, which is situated
about half a league to the south-east of Spa. The Sauvenière
water is saline, ferruginous, and acidulated. It is very sparkling,
containing nearly twice as much carbonic acid gas as the
Pouhon. The chief use made of the Sauvenière is in treatment of sterility in anæmic women, who frequently derive
extraordinary benefit from this water. By the side of the well
is the form of a shoe deeply engraved in the rock, and tradition
asserts, and ladies believe, that whoever quaffs a glass of the
Sauvenière standing, with her right foot in the "pied de St
Remacle," will within the year augment the population.

A few yards from the Sauvenière is the Groesbeck spring,

* "Lectures on the German Mineral Waters," by Siguismund Sutro, M.D.,
p. 327.

so named from some Baron de Groesbeck, who, very wisely foreseeing that posterity might otherwise ignore his existence, in 1771 put a tablet over the fountain, commenting on its virtues and his own. This water is less ferruginous and saline, though more gaseous, than any of the other springs.

The last of these mineral sources that I visited were the three *Tonnelets,* so named from the wooden receptacles in which the water was originally collected.

The "Tonnelets" are by far the most agreeable of the Spa waters, and one of them was not inaptly described by an old writer as being—" most grateful, subacid, vinous, smart, and sprightly, not unlike the briskest champagne wine, imparting exceedingly little, if any, vitriolic or ferruginous taste. Being drank, it generally sits lightly and agreeably on the stomach, and though exceedingly cold, it warms, cheers, and invigorates."*

* " An Essay on Waters," by C. Lucas, M.D. of Dublin, vol. ii. p. 199, London, 1756.

CHAPTER XXVII.

AIX-LA-CHAPELLE AND BORCETTE.

INTERESTING as are the reminiscences attaching to Aix-la-Chapelle, a very brief notice will suffice of this place as a modern spa.

Aix-la-Chapelle, or Aachen, is situated some thirty miles to the south of Spa, at the foot of a range of well-wooded hills of no great height. Like most of the watering-places, Aachen consists of an old town immediately around the wells, and a modern suburb in which the invalid visitors pitch their quarters, as far away as the extent of the place will allow of from the waters they have come to make use of. This latter portion occupies the upper part of the town on the Borcette or south side, extending from the theatre to the railway station ; and in it the streets are wide, clean, rectangular, dull, and uninteresting.

The springs are divided into upper and lower, of which the former are the hottest. The "Elisenbrunnen" is situated midway between the old and new town, close to the theatre, and issues forth under a handsome colonnade, where the water drinkers promenade, in the intervals between their potations, from five until eight o'clock in the morning, and again in the afternoon.

I am indebted to my friend Dr Velten for the following tables, which show that, in addition to the substances previously found in them, these springs contain iron, potassa, iodine, and bromine.

Composition of the Aix-la-Chapelle Sulphur Waters.

Not Volatile Ingredients.	Emperor's Spring.	Cornelius Spring.	Rose Spring.	Quirinus Spring.
Chloride of sodium, . . .	2·63940	2·46510	2·54588	2·59595
Bromide of sodium, . . .	0·00860	0·00360	0·00360	0·00360
Iodide of sodium, . . .	0·00051	0·00048	0·00049	0·00051
Sulphuret of sodium, . . .	0·00958	0·00544	0·00747	0·00234
Carbonate of soda, . . .	0·65040	0·49701	0·52926	0·55267
Sulphate of soda, . . .	0·28272	0·28664	0·28225	0·29202
Sulphate of potassa, . .	0·15445	0·15663	0·15400	0·15160
Carbonate of lime, . . .	0·15851	0·13178	0·18394	0·17180
Carbonate of magnesia, . .	0·05147	0·02493	0·02652	0·03346
Carbonate of iron, . . .	0·00955	0·00597	0·00597	0·00525
Silica, ...	0·06611	0·05971	0·05930	0·06204
Organic substance, . .	0·07517	0·09279	0·09151	0·09783
Carbonate of lithia, . . .	0·00029	0·00029	0·00029	0·00029
Carbonate of Strontian, . .	0·00022	0·00019	0·00027	0·00025
Sum of the non-volatile contents,	4·10190	3·73056	3·89075	3·96968

The medicinal effects of these waters depends, however, in a great measure, on the amount of sulphuretted hydrogen gas they contain, and which is here combined with a larger amount of nitrogen than in any other European sulphurous sources. From the preceding table we learn that there is no real difference between the springs, which only vary in temperature according to their position with respect to their common source.

There is a chalybeate spa in the Theater Strasse, with a bathing establishment attached to it, the water of which is cold, and is said to contain half a grain of iron to the pint; but the resident physician with whom I visited the springs of Aix-la-Chapelle attached little therapeutic importance to this source, which some consider as fictitious.

The mineral waters of Aix-la-Chapelle, though decidedly sulphurous, are seldom rejected by even the most fastidious stomach. Nay, strange as it seems that a fluid whose flavour is that of the washings of a dirty gun-barrel should ever be

L

palatable, I observed that after the first day or two the majority of the water drinkers at the Elisenquelle appeared positively to enjoy their matutinal potations.

The action of this spa is that of a stimulant, operating principally on the kidneys and skin. This determination to the surface and renal organs explains the efficacy of the water in many cutaneous diseases, glandular enlargements, biliary obstructions, atonic dyspepsia, renal complaints, uterine derangements, impaired health from metallic poisoning by mercury or lead, and in cases of lurking constitutional syphilis. The mineral waters of Aix-la-Chapelle are also prescribed in cases of chronic rheumatic-arthritis, rheumatism, and sciatica; and Dr Velten informs me that he has seen benefit derived from their use in some forms of chronic bronchitis and catarrh.

Fully two-thirds of the invalid visitors to Aachen, as far as I could judge, suffer from cutaneous eruptions, or from chronic rheumatism. In these diseases the internal use of the water is combined with baths, which, indeed, are the chief part of the treatment here. The principal baths are the "Bains de la Rose," in the Comphausbad Strasse, which are supplied from the lower source, and are well arranged and comfortable. The douche is peculiar; the attendant enters the water with the bather, and turns on the full force of the hot steam, through the hose, on the affected parts, kneading and rubbing them diligently with his disengaged hand at the same time. The temperature of these baths is 116°.

The course of the *Aachen* baths and waters is usually six weeks, and few can continue it longer, as the effect is so debilitating that most patients can only use the douche twice a week, and at the end of a course, even when cured of their original complaint, generally require a short course of some chalybeate water.

Before leaving Aix-la-Chapelle we visited the neighbouring springs of BORCETTE, or Burtscheid. This is a suburb of Aix, from which it is separated by the viaduct of the railway, and is a very quaint-looking old town, divided into an upper and lower quarter, consisting for the most part of dwellings of the poorer class, with the exception of eight or ten large bath-houses, a few hotels, and an ancient monastery of the ninth century, now the parish church, on the hill above the town. That we were in a place rich in thermal springs was evident the moment we entered Borcette, as through almost every street ran a river of hot mineral water.

The mineral springs of Borcette are divided into sulphurous and saline. The sources in the upper quarter of the town are distinguished from the lower springs, as well as from those of Aachan, by not containing either sulphate of soda or sulphuretted hydrogen gas. Of the sulphurous waters of Borcette, the most important is the Trinkquelle. The temperature of this spring is 138°, and it contains 30 grains of saline ingredients to the pint, 20 grains of which is chloride of sodium, 2 grains sulphate of soda, and 6 grains of carbonate of soda.

Amongst the non-sulphurous sources that most generally used is the Kochbrunnen, the temperature of which is 150°, and its chief saline constituents carbonate and sulphate of soda and chloride of sodium.

The waters of Borcette, which are warmer than any of those of Aix, are employed internally and externally, and are prescribed in cutaneous diseases—in dyspepsia and hepatic complaints, and in calculous affections. Moreover, the local physicians say they are applicable in some cases of scrofula, gout, and rheumatism. They are annually resorted to by close on twelve thousand visitors, chiefly Germans and French, this place being a much cheaper residence than Aix-la-Chapelle.

CHAPTER XXVIII.

EMS.

THE route from Aix-la-Chapelle to the next spa on our list led us through Cologne and along the Rhine, as far as Oberlandstein, whence twenty minutes' journey by train brought us to Ems. This town which, with the exception of Carlsbad, is the most aristocratic watering-place in Europe, is situated on the right bank of the Lahn, and is divided into " Bad-Ems," the quarter around the springs, and " Dorf-Ems," the adjoining suburb. Bad-Ems consists of a row of houses built in crescent shape, between the river and the precipitous hills, which rise immediately behind it, and is connected with the railway station by a handsome bridge. Opposite to this is the chief drinking spring and the cursaal, forming the extremity of the new quarter, from which an avenue of hotels and lodging-houses, fully a mile long, extends to Dorf-Ems, or the old village.

There are no less than twenty-five saline chalybeate springs here, of which, however, only three are used internally.

The Ems waters are strongly alkaline, clear and sparkling, and vary in their temperature from 118° to 83°. In taste, the hotter sources, to some extent, have the peculiar " chicken broth flavour " of the Kochbrunnen of Wiesbaden, conjoined with a slightly chalybeate taste.

Composition of the principal Mineral Springs of Ems.

	Kesselbrunnen.	Krähnchen.	Furstenbrunnen.
Temperature, . . .	116°	90°	96°
Carbonic acid, . . .	16·4480	26·8160	15·6760
Carbonate of soda, . .	14·7418	12·6108	16·5526
Carbonate of strontia, . .	trace.	trace.	trace.
Carbonate of lime, . .	1·4474	1·4400	1·5263
Carbonate of magnesia, .	0·3200	0·4975	0·6206
Carbonate of protoxide of iron	0·0576	0·0096	0·0195
Oxide of manganese, . .	trace.	trace.	trace.
Sulphate of soda, . .	0·3538	0·3981	0·3678
Chloride of sodium, . .	7·0216	6·3349	5·8335
Chloride of magnesia, . .	0·3318	0·3758	0·5248

The springs of Ems, being alkaline, saline, and alterative, are principally suited for cases requiring a mild, anti-acid aperient, increasing the secretions and improving the appetite. They are applicable in some chronic uterine ailments, and nervous diseases arising from these causes in females. They are, moreover, strongly recommended in aphonia, hooping-cough, dyspepsia, and many other complaints.

By some writers these waters have been advised in the treatment of chronic bronchitis, senile catarrh, the early stages of consumption, and "debility of the chest." As I do not understand the meaning of the latter term, I shall express no opinion on it. But with regard to the statement which appears in several books, English as well as foreign, on this subject, that cases of consumption may be benefited by the mineral springs of Ems, I must express my doubt that any case of phthisis is curable by any mineral water whatever.

CHAPTER XXIX.

SCHWALBACH AND SCHLANGENBAD.

FROM Wiesbaden, Schwalbach may be reached in a couple of hours by diligence, or by the Rhine steamers to Eltville, and thence by coach, in the same time. On my arrival at Schwalbach by the latter route, alighting at the Nassauer Hoff, I proceeded to call on the principal physician of the town, to whom I had a letter of introduction. Before accompanying Dr Genth, however, through the springs and baths, a brief *coup d'œil* over the town will not be misplaced.

Langen-Schwalbach is said to have been the favourite watering-place of the Roman legions stationed in Gaul. But it was almost unknown by English valetudinarians until forty-five years ago, when Sir Francis Head published his " Bubbles from the Brunnens," and such was the effect of his panegyric, that the hotels of Schwalbach are now crowded with English and American visitors during the season.

The town itself contains between two and three thousand resident inhabitants, and is situated in a little valley sunk between the lofty hills which form the table-land of the Taunus range. The name " Langen " is well applied, for the entire place consists of a long straggling street, expanding near the centre into a kind of open crescent, with two short diverging streets, the whole being built somewhat in the shape of a capital Y.

In the way of amusements Schwalbach has not much to

boast of; excellent bands play at stated hours in the Allée; there is a pretty park and miniature lake, reading and billiard rooms in the village, and charming walks and drives in the vicinity, *et voila tout.*

The principal mineral springs of Schwalbach are the " Weinbrunnen," the " Stahlbrunnen," and the " Paulinen-brunnen."

The Weinbrunnen is agreeable, piquant, and ferruginous in taste. This source was first brought into vogue by Dr Theodore of Worms, who in 1581 extolled its use as a remedy for the unpleasant feelings experienced the morning after an extra cup of wine, and hence is derived its name.

The Weinbrunnen is considerably stronger in iron than either the Stahl or Paulinenbrunnen, but containing less carbonic acid gas is less exciting, and agrees better with some constitutions than the other sources.

The Stahlbrunnen, which is situated at the extremity of a shady promenade,—the Allée,—is very similar to the Wein-brunnen, but is still more gaseous. The temperature of the water is 50°, or 1° higher than the last spring.

From the Stahlbrunnen a very pretty walk through the Allée, over which the foliage formed a complete arch the entire way, brought us to the Paulinenbrunnen, of which every reader of the "Bubbles from the Brunnen" will recollect Sir Francis Head's glowing description; but were Sir Francis now to revisit his favourite spring, he would hardly recognise in it his own picture. For ever since Dr Fenner's death, some years ago, the *Paulinen* has been declining in favour, and is now comparatively disused. Its temperature is 2° lower than the Stahlbrunnen. Being the mildest of the springs, it agrees in some cases in which the stronger waters of the Wein or Stahlbrunnen would not be tolerated.

In nearly all cases in which a ferruginous water is

required, the springs of Schwalbach have been recommended, and there is no Continental chalybeate spa more resorted to by English invalids.

In the following table I have nearly followed Dr Sigismund Sutro's analysis of the

Composition of the Schwalbach Mineral Sources.

Contents.	Wein-brunnen.	Stahl-brunnen.	Paulinen-brunnen.	Rosen-brunnen.
Carbonate of protoxide of iron, .	0·83	0·75	0·65	0·91
Carbonate of lime, . . .	2·11	1·45	2·95	2·95
,, ,, magnesia, . .	3·12	0·88	2·75	0·98
,, ,, soda, . . .	0·17	0·25	0·45	0·35
Chloride of sodium, . .	0·18	0·34	0·03	0·32
Sulphate of soda, . . .	0·16	0·21	0·02	...
Total solid contents, . .	6·59	3·83	6·86	5·51
Carbonic acid gas, . . .	26 cub. in.	28 cub. in.	39½ cub.in.	26 cub. in.

Three-fourths of the invalid visitors to Schwalbach suffer from some form of anæmia, amongst whom may be found a great many ladies labouring under chlorosis, or whose ailment is weakness and want of tone, following a London season, and for such patients the quiet of Schwalbach is perhaps not less useful than its chalybeate springs. In the debility following convalescence from severe and exhausting maladies, and in the weakness resulting from long-protracted nursing, the Paulinen or Weinbrunnen, sometimes even diluted, are often serviceable. Nearly all authorities on the subject speak of the powerful action of this spa in certain forms of functional derangement of the female system, and my own experience in several instances leads me to share this opinion.

In chronic indigestion, consequent loss of flesh, and consti-pation, depending on a relaxed and torpid condition of the mucous membrane of the stomach and alimentary canal, which the drastic remedies, that many people take habitually, can only aggravate—a visit to Schwalbach may be prescribed.

These waters should be taken fasting, in doses of from half a glass to two small glasses twice a day, and should be followed by a brisk walk. Their use would prove highly dangerous in cases of hæmorrhagic or organic disease of the lungs, heart, or kidneys.

SCHLANGENBAD lies six miles from Wiesbaden, midway between Eltville and the spa I have just described, in a valley almost hidden amongst thickly-wooded hills. It can hardly be called even a village, for it consists merely of a few bath-houses, hotels, and some twenty or thirty large barrack-like lodging-houses, irregularly scattered through the little valley. Being situated on the south-western slope of the Taunus range, and well protected from harsh winds by the hills, it enjoys a more genial climate than Schwalbach. There is in Schlangenbad a kind of " Sleepy-Hollow " atmosphere, which, irksome and even injurious as it would prove to many, must, I am sure, be most useful in some cases of nervous irritability and excitement, resulting from extreme mental tension and over attention to any absorbing pursuit in busy civic life.

Apart from this placid tranquillity, this *dolce far niente* existence, there is little or no attraction in Schlangenbad for any but real valetudinarians. For, beautiful as is the scenery, and interesting as are the excursions in its vicinity, they are almost equally accessible from Schwalbach or Wiesbaden.

The waters of Schlangenbad belong to the same class of mineral springs as those of Pfeffers and Wildbad, being, however, stronger than either of these spas. They are mildly alkaline and thermal, and are chiefly employed for bathing purposes. There are eight distinct springs, which vary in temperature from 77° to 90°. They rise at the foot of the adjacent mountain, whence they are conducted to the bath-houses. They all contain about eight grains of solid ingredients, with two cubic inches of carbonic acid gas, and the same amount of nitrogen in the pint. Of the solid constituents, about one-half consists of carbonate of soda, together with two grains of common salt, and one grain each of carbonate of lime and carbonate of magnesia. The special action of this spa seems to be on the skin, which it is said to render soft and white; and, therefore, I need hardly add, is largely patronized by the fair sex.

The power of diminishing nervous irritability has been ascribed to these waters by many writers, and they are, moreover, largely employed in the treatment of chronic rheumatism, as well as neuralgia and other nervous affections. In cutaneous diseases, too, such as lichen and prurigo, when a remedy is required to allay excessive irritation of the surface, a course of the Schlangenbad baths is often prescribed with advantage. Hufeland and other German writers recommend these baths in the articular rigidity of advancing years, as the veritable "Fountain of Youth" of the fairy tales. As a specimen of the rhapsodies which German spa physicians sometimes indulge in, I shall conclude this chapter with the following *morceau* from the late Dr Fenner of Schwalbach, who, in describing the effects of this bath, thus falls into an ecstasy of praise—" Vous sortez des eaux de Schlangenbad rejeuni comme un Phœnix—la jeunesse y devient plus belle, plus brillante, et l'age y trouve une nouvelle vigeur."

CHAPTER XXX.

WIESBADEN.

WIESBADEN enjoys a situation that renders it one of the most picturesque of the German spas, and which should also bless it with a climate superior to many of them, being built in an opening valley, extending from the southern slope of the Taunus Mountains to the Rhine, and thus completely protected from the north and east winds. The first view of the town is certainly prepossessing: the streets are wide and clean; the buildings are large, bright, and new looking; and the square in front of the cursaal, and that edifice itself, are really handsome. Since I first visited Wiesbaden, however, it has lost much of the life and gaiety which formerly were the special characteristics of this place. The political changes by which, nearly ten years ago, it was converted from the · flourishing capital of an independent state into a provincial watering-place, have probably more to do with the somewhat triste aspect of this once gayest of spas than the closure of the gambling tables, which has fortunately taken place more recently.

The hot springs of Wiesbaden were resorted to by invalids at a very early. date. Pliny describes them,—" Sunt et Mattiaci in Germania fontes calidi trans-Rhenum, quorum haustus triduo fervet; circa marginem pumicem faciunt aqua,"* and the remains of numerous "Balnearia" have been discovered in the vicinity of the mineral sources.

There are no less than twenty-two thermal springs in Wiesbaden. Some writers describe each of these separately, and assign different properties to each, but as all issue within an area of 3000 yards, and vary chiefly in temperature, in all probability they originate from the same source, and therefore, in the following remarks, I have confined myself to the Kochbrunnen, which I regard as a type of the other springs.

The Kochbrunnen rises nearly in the centre of the town. The appearance of the water is far from inviting, having a turbid yellowish colour, with a scum floating on the surface. The taste, however, is by no means unpleasant, and Sir Francis Head's comparison of it to "weak chicken broth" has been copied by every succeeding author as conveying an exact idea of its flavour. Its temperature is 155°, and the saline constituents in a pint of this water are 52 grains of common table salt, 4 grains of chloride of lime, 3½ grains of carbonate of lime, 1½ grains of carbonate of magnesia, and rather more than a grain of chloride of potassium. Besides these, it contains sixteen other ingredients, but in such small proportions as to be of no practical importance.

The thermal springs of Wiesbaden are principally employed for bathing; but as the Kochbrunnen is also extensively used internally, a few observations on its effects when employed in this way are necessary. This source is daily frequented by a considerable number of invalids, most of whom seem to suffer from gout, or dyspepsia, or hepatic disease.

Like all other mineral waters, this should be taken fasting, before breakfast. The dose varies from eighty to thirty ounces of the Kochbrunnen, which is sipped slowly, while the patients

* Pliny, "Hist. Nat.," lib. xxxi. c. 17.

promenade about the spring, and the exercise, I have little doubt, does almost as much good as the water.

The action of Wiesbaden mineral water is mildly purgative and diuretic. The circulation is always more or less excited by it, the biliary secretion is augmented, the action of the absorbents is quickened, and the appetite is sharpened, though this should not be indulged. If the Kochbrunnen be now persisted in for some weeks, the bulk of the body diminishes visibly, the expanded abdomen subsides, obesity disappears and the outlines of muscles, previously concealed by super-abundant fat, are thus brought into view.

After some time the blood generally becomes thinner and less rich in fibrinous compounds, respiration is now freer, the skin becomes clear and healthy-looking, and the valetudinarian experiences a general feeling of *bien etre*.

Such are the effects of the Kochbrunnen when it agrees with the patient. Unfortunately, however, it does not answer every case, though some of its panegyrists seem to think it does, being especially contra-indicated in cases of organic visceral and hæmorrhagic disease. When the course is prolonged beyond six or seven weeks, symptoms of febrile disturbance, or " saturation fever," are produced.

The diseases in which the internal use of the Wiesbaden springs are recommended are gout, dyspepsia, and plethora. In the irregular and atonic forms of gout, the saline waters of Wiesbaden often produce great benefit. Repeated experiments have shown that the amount of urea and uric acid eliminated by the kidneys is greatly increased under their use, and at the same time the patient's general health is improved by the alterative action of the spa. In some cases of lurking gout, this water brings on a fit of regular podagra, which is generally attended with complete relief of the other symptoms.

I have already mentioned that the warm springs of

Wiesbaden are principally used for bathing purposes, and the number of baths in the town is something extraordinary, amounting to nearly nine hundred. Almost every hotel has its thermal department, supplied from the twenty-five sources I have spoken of.

The bath should either be taken in the morning, fasting, or in the evening five or six hours after an early dinner. The patient at first should not remain more than ten minutes in the water, but may gradually, if so advised, increase this period to an hour.

The great majority of those who resort to Wiesbaden suffer from rheumatism, or from that combination of gout and rheumatism permanently affecting the joints which is known as chronic rheumatic-arthritis. Of one hundred and twenty-nine cases of chronic rheumatism treated by Dr Haas, in the hospital of Wiesbaden, by spa water, above thirty were completely cured, and seventy-nine were improved. The German physicians also recommend these baths as a remedy in rheumatic paralysis, where the power of voluntary motion is lost, and parts are forced into unnatural constrained positions, by long-continued arthritic inflammation.

CHAPTER XXXI.

HOMBURG, NAUHEIM, NEUENAHR, AND KREUZNACH.

THREE-QUARTERS of an hour's drive by the northern railway brought us from Frankfort to one of the most frequented spas of Europe, HOMBURG-ON-THE-HILL, which is situated on one of the lowest slopes of the Taunus range, immediately under the Great Feldberg peak. Of the three parallel avenues descending from the hill, of which, together with a couple of cross passages intersecting them, the town of Homburg consists, only one, the Luisen Strasse, deserves the name of a street. The upper part of this thoroughfare, with a few narrow lanes, forms the old town round the castle; while the lower part, which is entirely modern, contains the railway station, cursaal, and hotels, forming the new town.

The Kursaal occupies a prominent position in the main street, and is regarded by the inhabitants as the great feature of their town, being spoken of by them with that affectionate pride with which an Andalusian speaks of the Alhambra of Granada, or the mosque of Cordova, and in truth was, until a few years ago, one of the most elegant dens of iniquity in Germany, as well as the main support of that exalted potentate, the late and last landgrave of Hesse-Homburg.

The mineral springs of Homburg issue from a thick vein of quartz, covered by a stratum of gravel and clay, lying a hundred and fifty feet below the surface. These sources are all saline, their principal ingredient being common table salt, or chloride of sodium, in addition to which they contain a small proportion of iron. They also contain a large amount of free carbonic acid gas. All the springs, however much they may differ in strength, and even in composition, rise in close proximity to each other in the Kurgarten, a beautifully-kept park, connected with the cursaal, and occupying a small valley on the south-east side of the town.

The principal mineral spring is the Elisabethquelle, which was originally one of the sources of an abandoned brine manufactory, and had been disused for upwards of a century, until its medicinal properties were discovered in 1834. According to the late Baron Liebeg's analysis, each pint of this water contains 79 grains of common salt, 10 grains of carbonate of lime, 7 grains each of chlorides of magnesia and lime, 2 grains of carbonate of magnesia, half a grain of carbonate of iron, and traces of several other salts, which it would be useless for practical purposes to enumerate here. The quantity of carbonic acid gas the Elisabethquelle contains is its most remarkable characteristic, there being rather more than 48 cubic inches of gas in a pint of the water.

The Stahlbrunnen, which lies nearer to the cursaal, is an artesian well, and was opened about twenty-five years ago. The chemical composition of this spring is very similar to that of the Elisabethquelle, but it is not quite so rich in carbonic acid gas, and, as the name implies, contains somewhat more iron, amounting to nearly a grain to the pint.

The Kaiserbrunnen is also an artesian well containing a very large volume of carbonic acid gas, which gives rise once in every ninety seconds to a peculiar bubbling and ascent of the spring in the basin, like water on a quick fire. This is evidently caused by the subaqueous accumulation and subsequent escape of gas. In each pint of this spa are dissolved

117 grains of common salt, 13 grains of chloride of lime, 7 grains of chloride of magnesia, half a grain of carbonate of protoxide of iron, and a quarter of a grain of muriate of potash.

The next source is the Ludwigsbrunnen, which was the oldest used, and is the weakest of the Homburg springs.

The dose of Homburg water varies from six to forty-eight ounces, and must depend, in each particular case, upon the patient's age, sex, constitution, and ailment, as well as on the spring he may be advised to use ; the Kaiserbrunnen, for instance, being double the strength of any of the others. In general, however, small draughts of from four to eight ounces, repeated at short intervals, are more advisable than a larger quantity at one dose.

All the mineral sources of Homburg belong to the class of ferro-saline waters ; of which Kissingen, in Bavaria, and Cheltenham, in England, are examples, though of very different degrees of strength. In the treatment of gouty dyspepsia and hypochondriasis, attended with derangement of the liver, torpid or irregular intestinal action, and depression of spirits, a course of Homburg water often affords the best remedy. It is also used in cases of general plethora, and in what the German spa physicians term "abdominal plethora," an affection to which they attach considerable importance, and, so far as the cathartic properties of the water go, I have no doubt that it is very properly employed in such cases, especially as the "spa doctors," moreover, enjoin a regimen, which, of itself, would probably effect the cure.

The chloride of lime, contained by all these waters, increases the action of the absorbents and glandular system, and probably accounts for the benefits which have been occasionally observed to follow the use of this spa, in cases of scrofulous disease.

In varicose veins and ulcers the Homburg waters are said

M

to be productive of some advantage, by diminishing the tension of the diseased vessels, to which their chalybeate qualities impart a healthier tone.

In hysterical affections and amenorrhœa, this spa is frequently serviceable by restoring normal menstruation. In such cases it often proves the superiority of natural to artificial preparations, by the well-marked chalybeate action, which so small an amount of iron as that contained in these springs, the strongest of which holds only one grain of iron in a pint, is capable of exercising.

The usual duration of "the course" at Homburg is about three weeks, and I need not repeat the caution given in the introductory chapter against the use of these or any other mineral waters when symptoms of "saturation" have shown themselves, nor dwell on their possible dangers in organic and hæmorrhagic diseases. In combination with their internal use, the Homburg springs are sometimes employed for bathing, and the large amount of saline matter they contain renders these baths highly stimulating to the cutaneous capillaries. Their application in this way, however, is neither extensive nor generally advisable.

Nauheim is within an easy drive of Homburg, and is situated on the declivity of the Johaninsberg hill, on the railway from Frankfort to Cassel, and about an hour's journey from the latter place. The town is new and unfinished-looking, and contains a population of about 2000 inhabitants.

These waters have long been employed in the manufacture of salt, but only within the last few years have been resorted to medicinally. There were formerly a vast number of natural saline springs here, but their sources were gradually interfered with by the sinking of artesian wells, of which there are six or seven. The principal of these is the "Grosser-Sprudel," which burst from an unfinished artesian well, in

the night of December the twenty-first, 1846, in a thick
column of water to a height of nearly eighteen feet from the
surface, and thus it has continued to issue ever since. It is
enclosed within a large open stone basin, the water in which,
from the vast quantity of carbonic acid gas it contains, is
white with foam. It contains, besides, two hundred and
fifteen grains of saline matter to the pint; of this one
hundred and eighty-one grains are common salt, together with
fourteen grains of chloride of calcium, eleven grains of carbonate of lime, four grains of chloride of potassium, with other
salts, and a trace of bromine. Its temperature is about 90°.
This source is strongly purgative in doses of one glass, but is
principally used for bathing.

The Kurbrunnen and Salzbrunnen rise near each other.
The former is that usually employed internally; it is the
weakest of all the Nauheim springs, containing one hundred
and thirty-three grains of salt to the pint; in the chemical
composition of its ingredients it, however, resembles the
Grosser-Sprudel, as also does the Salzbrunnen, which contains
one hundred and sixty-nine grains, and the Kleiner-Sprudel,
containing one hundred and eighty-one grains, in the same
quantity of water.

The effect of these springs used internally is purgative and
diuretic. Their principal use is, however, for baths in cases
requiring a stimulating application to the skin, in obstinate
and languid cutaneous complaints. They are said to exercise
a specific action in accelerating and increasing the catamenia.
They are also prescribed in certain cases of chronic rheumatism,
in scrofulous tumours, and diseases of bone.

On the opposite bank of the Rhine to any of the German
watering-places just described are two spas, both of which
have of late years become of considerable importance; the
first of these is NEUENAHR, which may be reached by railway

from Cologne in an hour and a half, or still better by the Rhine Company's steamers to Remagen, from which a very pretty drive of barely seven miles along the valley of the Ahr brings us to Neuenahr. Midway on the road the spa tourist may stay for a moment at Ahrweiler to visit the source whence the Apollinaris water, so familiar on every table as a substitute for Seltzer water, is exported.

Neuenahr, which was first brought into notice in this country by the late Professor Miller of Edinburgh, belongs to the class of saline alkaline spas, and now attracts a considerable number of English and American gouty and dyspeptic valetudinarians. The place itself contains half a dozen good hotels, and the usual resources of a small German wateringplace, and needs no further description.

The chief saline ingredients of the Neuenahr springs are the bicarbonates of soda, magnesia, and lime, with a small amount of sulphate of soda, and a trace of protoxide of iron dissolved in a highly carbonated thermal water, the temperature of which is about 130°. It will be thus seen that this spa belongs to the same class as Vichy or Fachingen.

The Neuenahr springs are used internally in doses of from two to four small glassfuls twice in the day. The baths, however, and more especially the douche baths, the system of which is very perfect here, are the main attraction of this place for rheumatic and gouty patients, by whom, as well as by those suffering from chronic hepatic disease, dyspepsia, and some sub-acute, renal, and non-inflammatory uterine disorders, this spa is chiefly employed. The baths are also largely prescribed in chronic skin diseases, especially eczema and prurigo.

Nearly opposite to Schlangenbad, and about the same distance from the Rhine, is KREUZNACH, which is within less than half an hour's journey by railway from Bingen. This town, which contains a population of some 12,000 inhabitants, is prettily situated in the valley of the Nahe, and though long celebrated for its saline wells and salt-work, has only comparatively recently come into note as a watering-place, and is now one of the most important iodated-bromated spas in Europe.

The principal saline springs of Kreuznach are the Elisenquelle, which is the only source used internally, the Oranienquelle and the Nahequelle, which are used for baths, and locally in the form of concentrated brine or *mutterlange*. The Elisenquelle contains ninety-four grains of saline matter in each pint, and of this no less than seventy-two grains consist of common table salt, thirteen grains of chloride of lime, four grains of chloride of magnesia, and one and a half grains of carbonate of lime, the balance of the solid ingredients dissolved in this quantity of the water being made up of minute quantities of bromides and iodides of soda and magnesia, carbonates of baryta, and protoxides of magnesia and iron.

The "mutterlange," which is used as an addition to the ordinary baths as well as for local applications, is merely a concentrated solution of the salts just mentioned, and is obtained on a large scale as the residuum of the process by which table-salt is here manufactured.

This "mutterlange," one pint of which contains 1642 grains of the salts above named, is a thick, yellowish, oleaginous-looking fluid, intensely bitter in taste, and having a slightly iodated odour. The internal use of the water is generally conjoined with a course of the Kreuznach baths, and is administered in doses of from four to eight ounces twice daily. The duration of the baths varies from fifteen to forty-five minutes; and at first the pure mineral water is employed, but afterwards the "mutterlange" is usually added in increas-

ing quantities, commencing with a quart and gradually
reaching to ten or twelve quarts of the concentrated brine to
each bath. The usual duration of the course, if not interrupted by saturation fever or otherwise, is one month.

The Kreuznach waters and baths may be employed in the
treatment of every form of scrofulous disease, rickets, chronic,
glandular, and arthric enlargements, chronic inflammation or
congestion of the uterus and ovaries, the most frequent cause
of the anomalous chronic diseases of women, and more
especially of sterility, in cases of which I have repeatedly
prescribed the Kreuznach waters with remarkable effect. In
mammary enlargements, and especially in tumours causing
apprehension of cancer of the breast, I have seen the most
satisfactory results from a course of these baths, and the local
application of compresses soaked in the " mutterlange."

CHAPTER XXXII.

WEILBACH AND SODEN.

FROM Wiesbaden an hour's drive by train, through a richly-cultivated vine-country, and affording *en passant* a glimpse of the spires of Mayence, brought us to the little village of Flörsheim, where we alighted at 9 A.M. of a bright sunny morning. Hence, half an hour's drive through a continuous orchard conducted us to the baths of WEILBACH, which are situated on a slightly rising ground, in the valley of the Maine, midway between Mayence and Frankfort, and about two miles from the river. There is nothing in the shape of a town or even village at Weilbach, which merely consists of the bathing establishment.

The mineral source rises in the garden of this building, and is very, clear, of a slightly saline, alkaline, feebly sulphurous taste, but is not disagreeable, and has a tempera-ture of 55°. The sides of the basin are thickly coated by a soft, whitish, soapy deposit, which consists of carbonate of lime, sulphur, and a peculiar organic matter. The water, though not sparkling, contains six cubic inches of carbonic acid, and three inches of sulphurated hydrogen gases in the pint. According to the most recent analysis the same quan-tity of water contains four and a half grains of carbonate of soda, two and a half grains of carbonate of lime, two grains of carbonate of magnesia, two grains of common salt, one grain of chloride of magnesia, a little sulphate of soda, and some other ingredients, amounting in all to twelve and a half grains.

Used in full doses this spring is mildly aperient; in smaller quantities it is said to stimulate the appetite, and also to promote the removal of chronic visceral enlargements and congestions, especially of the liver. Most German physicians seem to be possessed by an extraordinary hallucination that all obscure diseases or complaints, the causes of which are not obvious to them (and I need hardly observe that the number is not limited), are produced by suppressed hæmorrhoids ; and it is further asserted by the local writers that the Weilbach springs are a panacea for these ailments. A large number of the invalids who visit this spa suffer from thoracic affections, and we are told that incipient consumption, chronic cough, and spitting of blood, are among the cases which come within the range of this remedy. I do not wish to express discredit of any particular writer; but I have no hesitation whatever in saying, as the result of very considerable experience of phthisis in various parts of the world, that I cannot believe that any consumptive patient was ever cured by the waters of Weilbach.

The next spa in my itinerary was SODEN, and thither I proceeded from Flörsheim by the Taunus railway, passing through luxuriant orchards, where the proprietors, gathering in the easy harvest, seemed to be keeping a holiday rather than engaged in the toil of husbandry.

Soden lies at the foot of the southern declivity of the Middle Taunus, below Königstein, and only twenty-five minutes' drive by rail from Frankfort. The village—the old part of which is built in a little valley between the hills—is protected from the north and east winds by the Taunus range.

The hotels are numerous and tolerably good. The soil is

peaty, and, as the only water found here is saline and mineral, the potable water is conducted from the neighbouring mountains. The saline sources issue from slaty rocks, and rise within a short distance of each other to the number of twenty-three.

The Soden springs are saline, acidulous, and ferruginous. In taste they are more or less piquant and saline, according to the amount of carbonic acid gas and chloride of soda which the sources all contain in different proportions: from twenty-four grains, which is the minimum in the " Milchbrunnen," to a hundred and twenty grains, which is the maximum in the " Salzquelle." The latter also contains the largest amount of iron and carbonate of lime.

The action of the Soden water is stimulant, diuretic, mildly aperient, and alterative. The resident physicians tell us that " the appetite invariably improves under the use of the water,—and this spa appears to be particularly well adapted to the treatment of atonic dyspepsia and languid digestion. The functions of the liver are excited by its use, and the biliary secretion is increased and becomes more healthy, therefore its use is indicated whenever this organ acts torpidly."

The Soden spa is also recommended in chronic mucous catarrh. This water moreover determines powerfully to the skin, and is used with great benefit in some obstinate chronic skin diseases, in scrofulous enlargement of the glands, and in chronic rheumatism.

CHAPTER XXXIII.

KISSINGEN, BOCKLET, AND BRÜCKENAU.

FROM Frankfort the journey by railway *via* Würzburg and
Schweinfurt to Kissingen now occupies rather less than six
hours. KISSINGEN, in the Bavarian department of the Lower
Maine, is situated on the River Saale, in the centre of a wide
valley surrounded by thickly-wooded hills. The town itself
is a mere watering-place, composed almost entirely of large,
newly-built hotels and lodging-houses, which during the
season generally accommodate about five thousand visitors.

The three principal springs rise close to each other in the
" Kurgarten," a kind of small park, a little to the south of the
town. The one we examined first was the Maxbrunnen
spring, which rises from the sandstone rock with a loud hissing
sound, and bright and sparkling, containing thirty-one cubic
inches of carbonic acid gas in each pint of the water. The
Pandur source is next to this, and is chiefly employed for
baths, but may be drank in many cases in which the Ragoczy
is too powerful. Like that spring, its action is aperient and
solvent, and also increases the cutaneous and renal secretions.

A few yards to the right of the last-described spring is
situated the most celebrated of these sources, and that which
is generally understood when " Kissingen Water" is spoken
of, namely, the Ragoczy, or Rakoczy. This spring was discovered in 1738, in the old bed of the river, which was then

turned into its present channel. Its taste is extremely unpleasant, being saline, bitter, acidulous, and somewhat chalybeate or astringent.

About a mile from the other springs, to the north of the town, rises the Soolsprudel, which intermits, or ebbs and flows, with great regularity, eight or nine times daily. If we arrive a little before the " flow," on looking down the shaft through the thick glass cap, we see the well apparently empty; we then hear a distant rumbling noise, which gradually becomes louder and draws nearer, till in about half an hour from the time it was first heard, the water is seen foaming below; it gradually ascends, and in another quarter of an hour reaches within a few inches of the top of the well, covered with white foam, and hissing and seething with great turbulence. Above the water hangs a heavy layer of carbonic acid gas, rising and falling with it. This gas is collected and ingeniously utilised in various forms of gas baths. The Soolsbrunnen acts as a powerful stimulant to the skin, and is principally used for bathing in the treatment of rheumatic, neuralgic, and cutaneous affections, and in scrofulous cases. The mother-lye, or concentrated brine of the Soolsprudel, is applied locally with great benefit in scrofulous and other glandular enlargements.

The Kissingen springs are used in almost every variety of dyspepsia. The spa physicians assert that they are the best remedy for chronic or habitual constipation, and they certainly act as very brisk and active cathartics, leaving no subsequent debility after their operation. Hypochondriasis, so intimately connected with irregular gastric or intestinal action, is said to be peculiarly under the influence of the curative action of these waters; and judging from the physiognomy of the visitors, half the invalids who drink the Ragoczy seem to labour under that malady.

Composition of Kissingen Mineral Springs.

	Ragoczy.	Pandur.	Max-brunnen.	Theresen-brunnen.	Sool sprudel.
Carbonic acid (cubic inch).	26·25	28·85	31·04	28·35	30·57
Chloride of sodium, . .	65·05	57·00	18·27	18·40	107·51
Chloride of potassium, .	0·91	0·25	1·00	0·85	0·97
Chloride of calcium,	...	...	...	...	3·99
Chloride of magnesia, .	6·85	5·85	3·10	2·75	24·51
Carbonate of soda, . .	0·82	0·03	0·38	0·39	...
Carbonate of lime, . .	3·55	5·85	2·59	2·00	1·65
Carbonate of magnesia, .	2·50	1·62	1·82	3·37	6·41
Carbonate of iron, . .	0·68	0·45	...	...	0·35
Bromide of sodium, . .	...	...	...	0·07	...
Bromide of magnesia, .	0·70	0·68	...	...	0·06
Sulphate of soda, .	2·00	1·75	1·86	1·35	25·30
Sulphate of lime, . .	2·50	0·75	0·65	0·75	...
Phosphate of soda, . .	0·17	0·05	0·12	...	...
Silica, ...	2·25	1·55	0·46	0·50	...
Oxide of aluminium,	0·18	0·05	...	...	...
Organic extract, . .	0·15	0·09	...	...	0·86
Loss, ...	0·38	0·37	0·38	...	...
Total solid contents in 16 ounces, .	85·74	76·39	30·65	29·63	187·68

In chronic enlargements and passive congestions of the liver and spleen the Ragoczy is often used with great advantage, but must be persevered in for some time after the patient, having gone through the ordinary course at Kissingen, has returned home.

In cutaneous affections, when connected with gastric derangement, these springs are considered to act as specifics. I know of hardly any mineral water which is not said to possess the property of curing sterility, and perhaps there is no place which has attracted so much attention on this account as Kissingen; and in cases where barrenness results from chronic ovarian, or uterine inflammation or congestion, I believe that the Kissingen waters may in many instances be prescribed with advantage.

A course of the Maxbrunnen is frequently conjoined with the use of the Pandur baths in the treatment of scrofulous cases. It is even asserted that the inhabitants of Kissingen who use the Maxbrunnen diathetically are therefore peculiarly exempt from strumous diseases.

From Kissingen a picturesque mountain drive of about fifteen miles leads the spa tourist to the important watering-place of BRÜCKENAU, which lies in the beautiful wooded valley of the Sinn. The springs are between two and three miles from the town, and around them are situated the baths, hotels, and lodging-houses, the royal palace, and Kurhaus. Two of the springs are on the left bank of the river, and contain only a small quantity of the salts of soda. One of these, the Sinnbergerquelle, is said to possess diaphoretic pro-perties, and is also prescribed in chronic bronchitis, scrofula, and calculous affections. On the opposite bank of the river is the Brückenauquelle, from which this place derives its reputation as a chalybeate spa. It rises under a pavilion, and flows into a basin incrusted with protoxide of iron. The taste of the water is strongly ferruginous, although it only contains a quarter of a grain of iron to the pint. But this is rendered most active by the immense quantity of carbonic acid gas, near thirty-eight cubic inches of which is combined with it.

The medicinal effect of Brückenau spa is tonic, and very stimulant. Its use is, therefore, dangerous whenever any disease, in which a powerful excitant would be improper, is present. It is, however, a very valuable stimulating chalybeate in certain cases, which I have already pointed out in the introduction, and need not repeat here.

Nearly five miles from Kissingen is the chalybeate spa of BOCKLET, a little village of lodging-houses grouped around a magnificent central "Kur-Haus," or bath-house. There are here several mineral springs, which are used internally and for bathing. All the sources are cold, chalybeate, and very gaseous. These waters contain about two-thirds of a grain of iron, together with twenty-seven grains of chloride of sodium, seven grains each of carbonate of lime and sulphate of soda, and other salts, amounting in all to nearly forty grains of saline matter to the pint. Their most important constituent, however, is carbonic acid gas, of which they contain thirty cubic inches to the pint; and it is this gas which confers its peculiar activity on the iron dissolved in the spa. As may be supposed from its composition, the Bocklet water is a very powerful stimulating saline chalybeate, and is peculiarly adapted to cases in which weakness of the digestive system is a prominent symptom.

CHAPTER XXXIV.

CARLSBAD.

CARLSBAD may be reached in a little more than two days from London, the most direct line of railway being that *via* Calais, Frankfort, Hoff, and Eger, when a short run of two hours by a branch line brings us into the most beautifully situated watering-place in Germany. It lies at the bottom of a deep ravine, intersected by the Tepel, and surrounded by lofty hills, wooded to their summits.

The resident population of Carlsbad does not probably exceed 4000 inhabitants, and the place may be described as consisting of a couple of streets called "Wiese," built along each side of the Tepel, and connected by eight or ten narrow bridges.

Most of the hotels frequented by foreigners, some three or four of which are excellent and not exorbitant in their charges, are situated on the right bank of the river, on which is also the Sprudel. The great majority of the invalid visitors to Carlsbad, however, take up their quarters during their course of the waters in some of the large lodging-houses on the Neue Wiese, or in the Mühlbad Strasse, where excellent apartments for a family may be had for from twenty to thirty florins a week, but must be engaged for at least a fortnight. Behind these a number of steep streets ascend the hill, and further on is a theatre and assembly rooms.

On the opposite bank is the Alte Wiese, a long range of small shops, extending southward along the Tepel Behind this is the Hirschen-Sprüng, whence, as the tradition asserts, a stag, hunted by Charles IV., sprang across the river and fell into the Sprudel, which is about half a mile distant, and by its piteous cries in the boiling water, attracted the emperor's attention, who, coming up, despatched the deer, and then refreshed himself with a warm bath, which had the effect of curing a bad leg, from which he suffered. This event was commemorated by order of the grateful monarch, by the name Carlsbad (or Charles's bath) being then affixed to the watering-place that at once sprang up around the emperor's bath.

The altitude renders the climate of Carlsbad remarkably changeable, and even in summer the morning air is often bitterly cold, especially when the north or east winds, to which this valley is exposed, predominate.

The town is built immediately over a vast subterranean boiler, covered in by deposits of the salts contained in the water, which evaporating has left them behind, and on the roof, or crust, thus formed, the town stands. The extent of this abyss or cavern of boiling water is quite unknown, as all attempts to fathom its depth have failed hitherto. The crust, however, that intervenes between the town and this cauldron is nowhere thicker than three feet, and in some places appears to be hardly as many inches. It therefore, I think, requires no great sagacity to predict, that in the event of any earthquake, the inhabitants of Carlsbad may probably try a longer bath in the Sprudel than any of its physicians would prescribe.

Carlsbad is too remote, dull, and expensive, to attract, like the rival spas of the Rhine, any but real invalids and their attendants.

The first morning I went out early to visit the springs I was surprised to see the crowd of people in the usually deserted-looking streets which lead to the wells. Every man, woman, and child carried a large beaker of steaming water in their hand, and were all gravely engaged in sucking the warm fluid through a small glass tube. Although this may be a very pleasant method of imbibing an iced sherry-cobbler on a hot summer's day, it certainly did not strike me as an agreeable mode of swallowing a large tumblerful of tepid Glauber salts and water at six o'clock on a damp autumn morning.

There are no less than nine thermal springs in Carlsbad which are used medicinally. These springs all rise from the same source, yet they differ materially in taste and temperature, according to the distance at which they issue from their common origin, and the nature of the strata they subsequently percolate. According to Mr H. Göttl's analysis, the following table shows the

Composition of the Sprudel.

Contents of one Pint of Water.	Grains.
Sulphate of soda,	14·9606
Sulphate of potash,	9·3696
Chloride of sodium,	8·7245
Carbonate of lime,	2·0198
Carbonate of magnesia,	0·3994
Carbonate of protoxide of iron,	0·0307
Aluminia,	0·2150
Silica,	0·0520
Total solid ingredients,	44·8340
Gaseous contents in Cubic Inches.	
Carbonic acid,	7·80337*
Nitrogen,	0·03181

The great spring to which Carlsbad owes its fame, as a watering-place, is the Sprudel, which is situated almost

* Dr Mannl calculates that 15,944 cwt. of Glauber salt, 13,000 cwt. of carbonate of soda, 10,000 cwt. of common salt, and 2500 cwt. of carbonate of lime are annually discharged into the river from the Sprudel, and totally lost.

N

exactly in the centre of the town, on the bank of the river. The water issues under a kind of open turret, through which the constantly-ascending cloud of steam that escapes is visible for miles around. Under this a large iron fountain is placed, from which the Sprudel issues in remarkable jets, resembling on a vast scale those that escape from a divided arterial trunk, but intermitting their pulsations, however, and varying in volume and height from eight inches to as many feet, the average height of the jets being about three or four feet. The metal basins are thickly incrusted with the white calcareous incrustations of "Sprudel-stone" deposited by the water, which issues at the temperature of 170°.

The other sources, which only differ from the Sprudel in their temperature, are the Hygieas-Quelle, temperature 165°; the Muhlbrunnen, 142°; the Marketbrunnen, 135°; the Bernardsbrunnen, 154°; the Schlossbrunnen, 129°; the Felsen-quelle, 129°; and the Theresienbrunnen, 130°.

The Theresienbrunnen is now the most resorted to of the Carlsbad springs, and is said to be more aperient though less exciting than the Sprudel. This source is especially in vogue with the gentler sex. When I first visited it at half-past six o'clock in the morning the fair valetudinarians were walking up and down in great numbers, sipping beakers of this fluid to the accompaniment of a tolerably good band. A number of white-coated Austrian officers, who were most unremitting in their attentions to the ladies, most of whom were, however, of a certain age, together with a few Russians, some long-haired Saxon students, and a couple of dozen Israelites, from Frankfort, made up the assembly. I should not omit to add that one corner of the promenade was, by common consent, left free to a gentleman in black, with a thick moustache, who, held brief colloquies in turn with most of the people present, and was evidently a brother professor of the healing art.

Of the long list of doctors who have extolled the efficacy of these waters, and their own skill, since Dr Payer of Elbogen first chronicled their virtues in 1522, none have achieved such celebrity as the Bohemian poet Lobkowitz, whose elegant Latin ode to the Sprudel has been translated into almost every European language. Lord Alvanley and the late Dr James Johnston have both rendered this ode into English; and a few lines from the latter version will suffice to show the spirit of the original,—

> "Sacred Font! flow on for ever,
> Health on mankind still bestow;
> If a virgin woo thee—give her
> Rosy cheeks and beauty's glow.
> If an old man—make him stronger;
> Suffering mortals soothe and save,
> Happier send them home, and younger,
> All who quaff thy fervid wave!"

Like most other mineral waters, Carlsbad has been extolled as the panacea for almost every disease in the interminable catalogue of human infirmities. We might indeed send the majority of our clients there, if the result could be ensured which Dr Porges tells us occurred to one of his patients, "who got a most flourishing look; nay, even his mental faculties were highly raised in consequence of this full recovery." *

The pathological effect that may be produced by the Carlsbad water is sufficiently important to demand some notice. A tendency to hæmorrhagic and congestive affections and vertigo has been frequently observed here among patients, and ascribed to the mineral water. hence it is hardly necessary to add that it is contraindicated in all organic cerebral, pulmonary, or cardiac diseases.

The special therapeutic action of Carlsbad water is in the

---

* "The Mineral Waters of Carlsbad, from a Homœopathic point of View," by Dr G. Porges, p. 137, Prague, 1874.

treatment of hepatic diseases. In enlargements, passive
congestion, torpidity, and other chronic diseases of the
liver, a course of the Sprudel is in many instances the
most useful remedy that can be prescribed. In some forms
of jaundice the rapidity of its action is remarkable, and its
power of dissolving and eliminating biliary calculi is not
less so. There is no spa so applicable in the majority of
cases of hypochondriasis as this, the curative effects of which
are explained by its purgative properties and its special action
on the liver in such cases.

The Carlsbad springs are beneficial in all affections dependent on an obstructed or torpid condition of the abdominal
viscera. They are, therefore, especially useful to valetudinarians of an obese habit of body, and, conjoined with low diet
and abundant exercise, will reduce corpulency more efficaciously, and far more safely, than the absurd system which,
under the fitting auspices of a London undertaker, has
recently been so productive of evil in this country.

In some affections dependent on an impoverished state of
the blood, the Carlsbad waters act more beneficially than any
of the strong ferruginous spas. This is probably owing to
their stimulant and deobstruent qualities, by which the constitution is prepared for the action of the very small quantity
of iron they contain.

The table d'hôtes at Carlsbad differ from those of almost all
other German watering-places, in the simplicity of the fare
and the paucity of the dishes. I have seen a newly arrived
guest who allowed a few dishes to pass untasted, reserving
himself for imaginary *plats* to come, surprised to find dinner
concluded before he had tasted anything but the soup. In
truth, invalids have here little temptation to gastronomic
excess, and perhaps this explains in some degree the good
effect of Carlsbad on dyspeptic and gouty patients.

CHAPTER XXXV.

MARIENBAD AND FRANZENSBAD.

MARIENBAD, which may be reached by the same route as the last described spa as far as Eger, from which it is only three-quarters of an hour's distance by train, is one of the prettiest of the continental spas. The town is quite modern, and consists entirely of hotels and lodging-houses built around three sides of an extensive park or garden, which contains most of the mineral sources. Nearly in the middle of this park is a very handsome church, built in the shape of a Greek basilica. This belongs to the monks of Töpl, who are the proprietors of the town.

The hotels being of enormous size, look quite out of proportion to the little town in which they stand, but are crowded to excess during the season. Where there are patients, there will, of course, be doctors; and this is the case in Marienbad, where a dozen physicians besides three surgeons, reside in the various hotels from May to September.

Marienbad, though one of the most beautifully-situated watering-places in Europe, must, I think, be a very dull residence for those who do not require its mineral waters. The Kursaal, at the extreme end of the park, is a large and handsome building, and besides this there are the usual assembly and concert rooms and a small theatre. Opposite to the windows of our hotel was the principal mineral spring —the Kreuzbrunnen, which, as is the case with all the

Marienbad springs, owes its therapeutic properties chiefly to the amount of sulphate of soda or Glauber's salts it contains.

Composition of Marienbad Mineral Springs.

	Kreuz-brunnen.	Karolinen-brunnen.	Ambrosius-brunnen.	Ferdinands-brunnen.
Sulphate of soda, . . .	36·11	2·79	1·86	38·53
Chloride of sodium, . .	11·16	0·82	1·64	15·39
Carbonate of soda, . .	7·13	0·20	1·66 ·	6·13
Carbonate of lime, . .	3·93	3·66	2·89	4·10
Carbonate of magnesia, . .	2·71	3·94	2·72	3·04
Carbonate of iron, . .	0·17	0·44	0·34	0·39
Carbonate of manganese,	0·03	...	...	0·09
Carbonate of lithia, . .	0·11	...	...	0·06
Silex, ...	0·38	0·46	0·48	0·66
Total solid contents in grains	66·	14½	10½	73·38
Carbonic acid gas, in cubic inches, . . .	8½	15·43	12·9	13¾

The Marienbad waters act directly upon the liver, the secretion of which is notably increased by their use. Their first effect is that of a saline aperient, not followed, however, by the debility attending the exhibition of other equally powerful remedies of that class. On the contrary, the carbonate of iron they all contain, though in such small proportions, produces a very decided tonic effect. The appetite is almost invariably sharpened by them ; the pulse is generally at first quickened, and the kidneys secrete more copiously under their influence.

The Kreuzbrunnen, which is cold, is generally administered with a sufficient quantity of warm water to bring the temperature of the draught up to 90°. The usual dose as an aperient is from three to four glasses of this mixture. The special action of the Kreuzbrunnen is to stimulate all the abdominal organs, especially the liver, to increased action ; hence its curative effects in cases of general plethora, dsypepsia, hypochondriasis, as well as in some hysterical and uterine affections.

The Karolinenbrunnen is the strongest tonic source in Marienbad, containing about half a grain of iron in a tumblerful of the water, as well as double the amount of carbonic acid gas found in the Kreuzbrunnen, and may be prescribed in most cases of general and local debility requiring a ferruginous tonic, as also may the Ambrosiusbrunnen.

The Marienquelle, the original spa of this place, is no longer used internally, being employed only for baths, which are here regarded merely as adjuncts to the use of the other springs. These baths almost invariably produce a powerful diuretic effect, and are prescribed in scrofulous glandular enlargements, chronic rheumatism, and torpidity of the liver and bowels, but cannot be used safely except under the supervision of a local physician.

Mud Baths are used in Marienbad, but not to the same extent as at Franzensbad, in the following account of which place they will be described.

Gasbäder, or gas baths, are also employed here, both locally and generally. In the latter form of bath the patient enters a square box, shaped exactly on the model of a Chinese pillory, covering the entire body except the head, which protrudes through a hole in the lid. Into this he is put, dressed in his ordinary habiliments; the gas is let in through a pipe in the bottom of the box, and presently, as the gas rushes in, a sensation of tingling and pricking is felt creeping up the legs, and gradually extends over the entire body. The " gas-bad," which owes its medicinal application to Dr Sturve of Dresden, has now risen into great vogue with German physicians and patients in the treatment of some diseases marked by general torpitude and vascular languor, suppressed menstrual or hæmorrhoidal discharges, and scrofulous ulcers.

From Marienbad a drive of twenty miles by railway,

passing through the historic town of Eger, brings us to Franzensbad, which is situated in the midst of a bog surrounded by bleak-looking mountains, and consists of four streets crossing each other at right angles, built on piles driven into the soft mud beneath. The principal street is the Kaiser-Strasse, a long boulevard planted with chestnut trees. This contains the chief hotels and bath-houses, and leads up to the Franzensquelle source.

There are four mineral springs here, the composition of which will be seen by a glance at the following table :—

Analysis of Franzensbad Mineral Waters.

	Franzensquelle (by Berzelius).	Salzquelle (Berzelius).	Wiesenquelle (Zembsch).	Louisenquelle (Trommsdorf).
Sulphate of soda, ...	24·50	21·52	25·65	21·41
Chloride of sodium, . . .	9·23	8·76	9·32	6·76
Carbonate of soda, ...	5·18	5·20	8·97	5·49
Carbonate of lithia, . . .	0·03	0·02	0·02	...
Carbonate of magnesia, . . .	0·67	0·79	0·61	...
Carbonate of lime, ...	1·89	1·41	1·37	1·60
Carbonate of protoxide of iron, .	0·23	0·07	0·13	0·32
Carbonate of protoxide of manganese,	0·04	0·01	0·02	...
Phosphate of lime, . . .	0·02	0·02	0·02	...
Other salts, ...	0·47	0·49	0·46	0·22
Total solid contents (grains), .	42·18	38·29	46·58	35·80
Carbonic acid gas (inches), . .	40	26·88	30·89	32·54

In their action, the Franzensbad waters resemble those of Marienbad, but, containing less sulphate of soda and more carbonic acid gas and iron, are not so lowering. The use of these springs is said to produce a sedative action on the nervous system, while imparting strength and tone to the muscles ; they also purify the blood by their purgative and diuretic action, and improve its composition by the additional nutriment which the patient is now enabled to digest.

Therefore Franzensbad is largely resorted to by dyspeptic and hypochondriacal patients.

Near the Franzensquelle is the Gasquelle or gas source, over which baths have been erected; the effects of which, being precisely the same as those of the gas-baths of Marienbad, need not be again described.

The mud-baths are the special advantages of Franzensbad, and are extolled as a cure for every disease under the sun, *et quibusdam aliis.* The soft boggy earth which surrounds Franzensbad on all sides is the material of these baths. It is dug up and repeatedly forced through sieves, until it is perfectly free from all foreign matters, woody fibres, &c., and when it has attained a perfectly soft, homogeneous condition, is diluted into a semi-fluid, black, pultaceous mass, exhaling a strongly sulphurous smell, with the Louisenquelle water, heated to about 100°. Into this uninviting-looking bath the patient enters, and so dense is it, that it is generally some time before he can immerse his whole body. The use of these baths is by no means so unpleasant as their appearance, and the bather generally leaves with reluctance at the end of the quarter of an hour which is their usual duration; and then is placed in a plain tepid water bath, where he finds sufficient occupation for half an hour in restoring himself to something like cleanliness.

The principal saline contents of this mud are sulphate of soda, lime, magnesia, iron, and aluminia, silica, tannin, sand, resinous and vegetable matters.

The primary action of these baths is stimulant and exciting to the nervous system. They produce some degree of cutaneous irritation; whilst in them the skin looks corrugated and wrinkled, but feels smooth and glossy immediately after emersion. The appetite is almost always increased by the external use of this mineralised mud.

The *Moorbäder*, as these mud-baths are called, are used in chronic arthritic and rheumatic affections, in skin diseases of an obstinate, languid character; in similar ulcers; in glandular swellings, in paralytic complaints, particularly of the lower extremities; and are renowned for the cure of old and painful wounds.

CHAPTER XXXVI.

TEPLITZ, BILIN, PÜLLNA, AND SEDLITZ.

TEPLITZ, which, next to Carlsbad, is the most important of the Bohemian spas, may be reached by railway with equal facility from Northern and Southern Germany, being only sixty-three miles from Dresden, and sixty from Prague.

The town lies in a narrow and very fertile valley, well protected by the Erzgebirge mountains. Like the other Bohemian watering-places, Teplitz contains little, with the exception of its mineral springs, to attract tourists, being in all other respects a quiet little country town of about 6000 inhabitants. The Bad-Platz, containing the chateau, is a handsome park surrounded by hotels, lodging-houses, assembly rooms, theatre, and bathing establishments.

In the immediate vicinity of the town, and in the adjoining village of Schönau, there are no less than seventeen mineral thermal springs, differing only in temperature.

These springs are all saline, alkaline, and slightly chalybeate, but are none of them of any chemical strength sufficient to explain their undoubtedly powerful medicinal effects, which in great measure must be accounted for by the high temperature at which they are employed. The solid ingredients are about five grains to the pint of water, and chiefly consist of carbonate of soda.

Subjoined is the analysis of the principal spring of Teplitz, as given by Dr Sutro,* whose table differs somewhat from M. Wolf's as cited in Dr Seegen's work on mineral waters :†—

---

* " Lectures on the German Mineral Waters," by Sigismund Sutro, M.D., p. 25.

† "Handbuch der Heilquellenlehre," Von Dr Josef Seegen, p. 663.

Composition of the Hauptquelle.

	Grains.
Sulphate of potash, ...	0·43
Carbonate of soda, ...	2·68
Carbonate of lithia, ...	0·01
Carbonate of lime, ...	0·32
Carbonate of strontia, ...	0·01
Carbonate of manganese, . . .	0·08
Carbonate of magnesia, . . .	0·05
Carbonate of iron, ...	0·03
Chloride of sodium, ...	0·43
Chloride of potassium, . . .	0·10
Iodide of potassium, ...	0·05
Phosphate of aluminia, . . .	0·02
Silico-fluoride of sodium, . . .	0·13
Silica, ...	0·31
Crenic acid, ...	0·09

Total solid ingredients in 16 ounces of water　4·84

Notwithstanding the similarity of their chemical composition, the thermal waters of Teplitz differ so much in their therapeutic effects, that it would be quite unsafe to use any of them without first consulting a local practitioner as to the proper spring to be employed in each case. The principal source used internally is the Gartenquelle, which is a mild aperient and resolvent, said to be very efficacious in the treatment of chronic glandular and visceral enlargements.

The principal use of the Teplitz springs is in the baths, which generally occasion a good deal of vascular excitement, or slight febrile disturbance, and after some days' employment commonly produce a red cutaneous eruption with great irritation of the skin.

The Teplitz baths are used in the treatment of chronic rheumatic-arthritis and diseases of the joints, in spinal curvature and hip-joint disease, in cases of amenorrhœa and hysteria, in certain chronic skin diseases, in enlargements of the liver or spleen, hypochondriasis, and other chronic maladies in which a remedy which combines stimulant with tonic and resolvent properties is required.

I need hardly add that these waters are especially contra-indicated in all acute inflammatory or hæmorrhagic diseases.

Within nine miles of Teplitz, on the road to Prague, is Bilin, the so-called "Vichy of Germany." This little market-town of 3000 inhabitants lies about a mile from the remarkable basaltic mountain of the Biliner-Stein, and needs no descrip-tion here, being seldom resorted to by invalids, although its waters are largely exported.

The chief ingredient of the mineral water of Bilin is carbonate of soda, of which it contains twenty-three grains in the pint, being five grains stronger than the Fachingen spa, to which it otherwise bears a close resemblance. The chief source is the Josefsquelle, the analysis of which, according to Dr Redtenbacher, is as follows :—

Composition of Bilin Mineral Water.

	Grains.
Carbonate of soda, ...	23·106
Carbonate of lime, ...	3·089
Carbonate of magnesia, ...	1·098
Carbonate of protoxide of iron, . . .	0·080
Carbonate of lithia, ...	0·110
Sulphate of potash, ...	0·985
Sulphate of soda, ...	6·350
Chloride of soda, ...	2·935
Basic phosphate of aluminia, . . .	0·065
Silica, ...	0·244
Total solid ingredients, . .	38·062

The Bilin springs are employed in cases requiring an alkaline carbonated water, in diseases of the urinary organs and kidneys, in Bright's disease, in certain cases of gout, in jaundice, and in rheumatic affections of the joints.

Twelve miles from Teplitz are the sources of the well-known bitter waters of Püllna, Saidschütz, and Sedlitz.

The village of Püllna, on the road to Carlsbad, is the chief source whence the "Bitterwasser" is exported to every part of the civilised globe ; but as the water is very seldom drank on the spot, there is hardly any accommodation for invalid residents, beyond a very second-rate inn. The mode of collecting the "Bitterwasser" at Püllna is similar to that employed at Sedlitz and Saidschütz, and may here be described once for all. These three sources are situated in an extensive marly plain, the soil of which for a limited area round each spring has a peculiar light yellowish colour, and is perfectly destitute of vegetation. In this marly clay, wells or tanks are dug, and the rainfall and oozing of the soil is suffered to accumulate in them for months, dissolving out the soluble saline ingredients from the subjacent formations.

To return to Püllna, according to Professor Ticinus, the following is the analysis of

Püllna Bitter Water.

		Grains.
Sulphate of magnesia,	. . . .	96·975
Sulphate of potash,	. . . .	82·720
Sulphate of soda,	. . . .	123·80
Chloride of magnesium,	. . .	19·120
Nitrate of magnesia,	. . . .	4·602
Carbonate of magnesia,	. . .	6·280
Sulphate of lime,	. . . .	0·800
Carbonate of lime,	. . . .	0·760
Bromide of magnesium,	. . .	0·588
Phosphate of soda,	. . . .	0·290
Total solid ingredients,	. .	222·900

Together with 49 cub. in. of carbonic acid gas, in 16 ounces of the water.

To the eastward of the last-described source in the same plain lies SEDLITZ, or Seidlitz, whose name is the most familiar, and whose waters are the least used of all the German mineral springs. Sedlitz is a wretched-looking place, hardly meriting the name of a village, and the wells, whence the water should be derived, are a few shallow circular pits, whose contents very seldom find their way to this country. The actual Sedlitz water differs in every respect from the " Genuine Sedlitz powder " of our chemists. Instead of the cooling agreeable draught composed of tartrate of soda and potash and bicarbonate of soda, set into effervescence with tartaric acid, used in England under this name, the true Sedlitz-wasser is a bitter, nauseous, yellowish-looking fluid, the composition of which is not less different from that of its English namesake than its taste, the following being the ingredients contained in 16 ounces of the

Sedlitz Water.

	Grains.
Sulphate of magnesia, . . .	104
Sulphate of lime, ...	8
Carbonate of lime, ...	8
Chloride of soda, ...	3
Carbonate of magnesia, . . .	3
Total solid contents, . . .	126

With three and a half cubic inches of carbonic acid gas.

Half an hour's walk from Sedlitz, on a slight elevation above the plain, is SAIDSCHUTZ, the most largely exported of these waters. It is considerably stronger than the Sedlitz, and, according to Berzelius, the following is the analysis of 16 ounces of

Saidschütz Water.

	Grains.
Sulphate of magnesia, ...	84·16
Nitrate of magnesia, ...	25·17
Carbonate of magnesia, ...	4·98
Chloride of magnesium, ...	2·16
Sulphate of potash, ...	4·09
Sulphate of soda, ...	46·80
Sulphate of lime, ...	10·07
Oxides of manganese, iron, tin and copper, &c., .	0·28
Total solid constituents, . . .	178·77

These "Bitter Waters," are aperient, resolvent, and diuretic, varying in strength from the Püllna, which is the most powerful, to the Sedlitz, which is the weakest. They are all too strong, in general, to be used undiluted, and their action is quickened as well as rendered safer by mixture with an equal amount of warm water. They closely resemble the Friedrichshall spring, and are employed in the same class of cases—namely, in habitual torpidity of the intestinal canal, in congestions, torpidity, or enlargement of the liver and spleen, in plethora, in tendency to apoplexy and congestion of the brain, and similar diseases in which depletion is indicated. I have prescribed them extensively in the treatment of habitual constipation and hypochondriasis, and also more especially in the chronic diseases peculiar to women, in lieu of the drastic pills or draughts such patients seem so fond of taking. The usual dose is half a large tumblerful of the bitterwasser, with an equal quantity of hot water every morning, and may be repeated at intervals of two hours until the required effect is produced.

CHAPTER XXXVII.

GASTEIN, ISCHL, AND BADEN NEAR VIENNA.

WITHIN the same empire as the spas last described are two of
the most important health resorts on the Continent, viz.,
GASTEIN and ISCHL, both situated in the midst of the
Noric mountains in Upper Austria. The first-named watering-place may be reached in three days and a half from
London, *via* Munich and Salzburg. From Salzburg a drive of
some fifty miles, occupying about fourteen hours by diligence
or carriage, through magnificent Alpine scenery, and through
the wild and sombre defile of the Klamn Pass, brings us into
Wildbad-Gastein.

This village, if it may be so called, is one of the most exquisitely situated, but most primitive-looking, watering-places
in Europe. It is built at an elevation of upwards of 3000 feet
above the sea, on the side of a steep mountain, along which
some quaint ancient wooden houses, a long covered gallery,
known as the "Wandelbahn," that serves as a cursaal, together with a few modern stone buildings and hotels, are
thinly scattered. Immediately below the village is a deep
ravine, through which a mountain torrent, the River Ache,
rushes impetuously, forming two fine waterfalls, the incessant
din of which is one of the most noticeable features of this
place.

The warm springs of Gastein, some six or seven in number,
belong to the class of chemically indifferent thermal waters,

o

and vary in warmth from 115°, which is the temperature of the
Haupt-quelle, to 118°, which is that of the Spital-quelle. Dr
Pröll of Gastein enumerates no less than fourteen salts which
have been discovered in these waters.* But as the sum total
of all these saline ingredients is only 2½ grains in each pint of
the water, it will be sufficient to mention that the chief of
these is sulphate of soda with traces of chloride of sodium,
sulphate of potash and carbonate of the protoxide of iron.
It is obvious that such homœopathic quantities of any of these
constituents as is contained in the ordinary dose of the
Gastein water is quite insufficient to account for the undoubted
powerful therapeutic effects that occasionally follow a course
of these waters and baths. But without seeking to explain
the *modus operandi* of this spa, we may briefly mention the
cases in the Gastein baths, and the internal use of the waters,
are recommended. Having before pointed out the cases in
which other spas of the same class are employed, I need only
state in addition to those ailments in which Schlangenbad,
Wildbad, or Pfeffers are prescribed, namely rheumatic gout
and chronic rheumatism, contractions of the joints, neuralgia,
hysteria, and hypochondriasis, the pure atmosphere of this
charming mountain sanatorium renders Gastein peculiarly
suitable for valetudinarians who require not only a course of
the baths or waters, but also the tonic and invigorating
influence of the bracing climate.

ISCHL, the favourite health resort of the Emperor of Austria,
and hence the most fashionable and expensive of the Austro-
Hungarian spas, is situated in the very centre of that vast and
beautiful imperial domain generally known as the Salzkammergut, which intervenes between the confines of Styria
and Salzburg. This watering-place may be reached by

* " Gastein, Erfahrungen und Studien," Von Dr Gustav Pröll, " Brunneartz
in Bad. Gastein," p. 88.

railway from Vienna to Gmunden in nine hours, and thence by carriage in three hours. The saline springs of Ischl have been well known and largely employed for the manufacture of table salt ever since the commencement of the twelfth century, but only within the last fifty years have they been applied to medicinal purposes, and hence the handsome town of some 5000 inhabitants, well provided with hotels, casino, theatre, and all the ordinary attractions of a fashionable German watering-place, is entirely modern.

The baths of Ischl are divided into saline brine baths, or "Soolbäder," and saline vapour baths, or "Soolendampf-bäder." The former are made by the addition of the strong brine from the adjacent salt mines to a sufficient quantity of warm water. This brine or "soole" consists of two parts of the Hallstadt spring, with one part of the Ischl saline water, and is added in quantities varying from ten to thirty pints of brine to the bath, which is heated to about 90°. According to Professor Schrotter, the chief ingredients of one pint of this mixed brine or "Ischl-bade-soole," are 223 grains of common table salt, with traces of chloride of magnesia, bromide of magnesia, and sulphates of soda, potash, and lime.* The principal use of the Ischl baths is in the treatment of scrofulous diseases, rickets, chronic rheumatism, amenorrhœa, and some instances of sterility dependent on defective ovarian action, suppressed hæmorrhoids, certain cases of hysteria and hypo-chondriasis.

Within three quarters of an hour's journey from Vienna by the Trieste Railway, is the ancient watering-place of BADEN-BEI-WIEN, which is prettily situated on the Schwe-chat, at the foot of the Wienerwald Mountains. This spa is a favourite holiday resort of the Vienese bourgeois; and many of the baths, such as the "Herzogsbad," are vast

* "Ischl sur le Report Medical," &c., par le Dr J. Pollak, p. 56.

reservoirs of warm mineral water, capable of containing a couple of hundred bathers at a time, who, clad in a becoming costume, and surrounded by the crowds of spectators that fill the galleries above the baths, appear to seek amusement fully as much as health. But, though thus employed by those who do not require them, the thermal baths of Baden have been long esteemed as active therapeutic agents. These springs, of which there are here thirteen, were known to the Romans as the *Thermæ Pannonicæ*, and vary in temperature from 98°, the Josefsquelle, to 79°, the Peregrinusquelle. The chief saline ingredient of the Baden waters is sulphate of lime, of which they contain 5 grains to the pint, together with 2 grains of sulphate of soda, smaller quantities of the chlorides of magnesia, soda, and other salts, amounting in all to 14 grains of saline matter, and 2 cubic inches of carbonic acid, sulphuretted hydrogen, and nitrogen gas. The principal use of these baths is in the treatment of chronic rheumatism and rheumatic arthritis, scrofulous glandular enlargements and chronic skin diseases, especially those connected with the scrofulous diathesis.

CHAPTER XXXVIII.

CANNSTATT AND WILDBAD.

WURTEMBERG possesses two watering-places of considerable
importance. The first of these is Cannstatt, on the Neckar,
about three miles from Stuttgart.

There are no less than eighteen or twenty saline chalybeate
springs in Cannstatt.

Analysis of the Principal Springs of Cannstatt (Fehling).

	Sulzerrain-quelle.	Fräsnerische-quelle.	Sprudel.	Neuquelle, No. 1.	Neuquelle, No. 2.
Chloride of sodium, . .	16·29	19·50	16·42	12·63	7·59
Chloride of potassium, .	...	0·25	...	0·87	0·57
Chloride of magnesium,	...	0·18	...	...	...
Carbonate of lime, .	7·89	7·38	8·82	7·95	6·40
Carbonate of magnesia, .	...	0·31	...	...	...
Carbonate of protoxide of iron,	0·16	0·25	0·18	0·17	0·02
Sulphate of soda, . .	2·92	4·75	2·18	0·87	1·04
Sulphate of magnesia, .	3·53	2·25	3·61	3·89	3·34
Sulphate of lime, . .	6·43	7·75	6·32	6·88	5·06
Sulphate of potash, . .	1·23	...	1·38	...	..
All other ingredients, . .	0·16	...	0·17	0·09	0·08
Total solid contents, .	38·61	42·62	38·98	33·45	24·10
Carbonic acid gas,	23·5	19·4	27·7	14·6	8·8

The Cannstatt springs owe their efficacy to the combination
of different purgative salts, together with a small quantity of
iron, rendered peculiarly active and soluble by an excess of
carbonic acid gas. Thus they combine aperient with slightly

tonic properties. They are, therefore, valuable deobstruent remedies, and are also frequently prescribed by the physicians of Stuttgart in dyspeptic cases. The principal use, however, of these springs, is in cases where a mild tonic is indicated in chronic catarrhal affections of the mucous membrane, and in some scrofulous diseases.

The " Sprudel " spring is that most frequently prescribed. There is another celebrated source which is used principally for bathing, " Die obere Sulzquelle," a small pond of about half an acre diameter, formed by several springs. The water is so gaseous that it seems absolutely boiling, so hissing and bubbling is it with nitrogen and carbonic acid gases. The temperature is about 66°, and its medicinal use is chalybeate and solvent, being especially employed, in the form of local douche baths, in the treatment of catarrhal affections of the utero-vaginal and vesico-rectal mucous membranes.

WILDBAD, the most romantically-situated of the German watering-places, is beautifully placed in the very centre of the Black Forest, about six hours' journey by railway from Stuttgardt, and four hours' from Carlsruhe. From Pforzheim the branch line to Wildbad passes through the valley of the Enz, the narrow strip of land between which and the forest is cultivated with a care which throws Mr Mechi's model farming completely into the shade. Through the midst of this the Enz, here the noisiest and most turbulent mountain stream of its size that can be imagined, rushes white with foam, and bearing rafts so narrow, although longer than the " Great Eastern," that they hardly afford footing to the man who navigates them, down to the distant Rhine; while back from the river extend for miles on either side the gloomy pine-shades of the Black Forest.

On our arrival at Wildbad, having secured our rooms at the Hotel Klumpp,—which, I may remark, is one of the most comfortable hostelries in Germany, though there are several others, perhaps equally good, here,—I called on Dr Haussmann, to whom I had letters of introduction, and to his courtesy and assistance I am indebted for much of the information I obtained concerning this watering-place.

Wildbad, which stands in the narrowest part of the valley of the Enz, contains a population of about 3500 inhabitants. The Grand Bath House or Curhaus, which is opposite to the Hotel Klumpp, is the most perfect bath establishment in Europe for its size. Of its extent some idea may be formed from the fact that twelve hundred baths can be daily administered, each bath being, moreover, of considerable duration. The foundation of this building is cut out of the solid granite rock, through which the water percolates, and on which the baths lie with the intervention of a thin layer of fine sand. The establishment is equally apportioned to male and female baths. In both are large public "piscinæ," and smaller cabinets for those who prefer bathing separately. One characteristic of all these baths is the great height of the rooms, so that one is not plunged into a hot vapour bath before entering the water, as is the case in almost every other bathing establishment in Germany that I have visited. The water is remarkably clear, so that every grain of sand at the bottom is distinctly visible, although covered by some three feet of water, through which minute bubbles of gas are continually ascending. The temperature in the principal piscina is 96°, and in the other baths it varies from 92 to 103°.

The arrangements of this establishment are excellent; and the precautions taken to prevent those afflicted with contagious and horrifying diseases from bathing in the public baths, are especially deserving of imitation in every similar institution.

The reputation of Wildbad as a spa is of very ancient date,

and in the reign of Charles V., by the gratitude of some
courtier who here regained his health, a curious charter
was obtained for this town, which contained a proviso that
"all criminals, with the exception of murderers and highway
robbers, might here enjoy peace and quiet undisturbed, for a
year and a day."

According to Fehling, the following is the

Analysis of Wildbad Water.

		Grains.
Sulphate of soda,	. . . .	0·29
Sulphate of potash,	. . . .	0·10
Chloride of sodium,	. . . .	1·80
Carbonate of soda,	. . . .	0·83
Carbonate of magnesia,	. . . .	0·07
Carbonate of lime,	. . . .	0·73
Carbonates of iron and manganese,	. .	0·02
Silex,	. . . . . .	0·48
Total,	. . . .	4·35*

The gaseous constituents of this water have probably more
to do with its therapeutic effects than its saline ingredients,
and are very abundant, 100 parts consisting of 91·56 of
nitrogen, 6·54 oxygen, and 2·00 carbonic acid gas.

These constitutents, solid or gaseous, are, however, insufficient to account for the active therapeutic properties of
Wildbad water. Some writers ascribe these solely to its
temperature, which is 98°, or exactly that of the blood.

With respect to the *modus operandi* of these springs, Dr
Haussmann's theory is, that the Wildbad water contains no
lime whatever, although it is a very powerful solvent of that
base, and therefore that it acts by dissolving and removing the
salts of calcium which exist in excess in the blood and tissues.
He also gave it as his opinion that by increasing the fluidity
of the blood it thus facilitates the elimination of morbid
materials from the system, and in part also obtains the same

* "Handbuch der Heilquellenlehre," p. 652.

result by stimulating the excretory organs, especially the kidneys and skin.

The first effect of the Wildbad baths is a peculiar sense of comfort, or *bien-être*, which has been, I think, somewhat exaggerated by most writers. Succeeding to this is a slightly stimulant or exhilarating influence, which, if the bath be too long continued, is followed by a feeling of lassitude and a soporific tendency. Therefore the local physicians enjoin their patients to commence with a bath of ten minutes' duration, which may be gradually increased until it at last reaches half an hour's immersion, beyond which it will be very seldom, if ever, proper to prolong the bath. After some time the patient will generally experience the symptoms of what has been already described as the spa fever, or saturation point, and this usually proves critical, and is a precursor of the cure of the ailment for which the invalid has visited Wildbad.

These baths are largely employed in the treatment of neuralgia and sciatica, in some of the diseases peculiar to women which are connected with chronic uterine or ovarian inflammation, also in cases of functional amenorrhœa; in various scrofulous glandular affections, and above all in rheumatic gout and chronic rheumatism. In the last-named complaint it is indeed that Wildbad seems to exercise its most marked curative effect; more especially is this sanative influence shown in cases of chronic rheumatic-arthritis, in which the action of the joint is impaired, or even its form altered, by morbid deposits. Such structural changes are often rectified, and the effused matter absorbed by the combined internal and external use of this, apparently, simple water, when more pretentious remedies have been long tried in vain.

CHAPTER XXXIX.

BADEN-BADEN.

THE " Queen of the Spas," as Baden has been designated, is but six hours from the watering-place described in the last chapter, and within thirty-two hours' journey from London. The situation of Baden, on the declivity of the Schlossberg overlooking the valley of the Oosbach, is exquisite ; the hotels and lodging-houses are commodious ; the people are civil ; the shops are good ; the resources for the amusement of visitors are numerous ; and the mineral springs, which are recommended in almost all chronic diseases, may be found really useful in some cases.

Baden is divided into two distinct towns—the old and the new,—the buildings of which are as different as their populations. The former, which is situated on the hill, consists of narrow lanes of quaint, old-fashioned houses, rising in successive terraces, is peopled by the indigenous inhabitants, and contains the mineral sources. The new town occupies the valley on the south side of the river, and contains the " Trinkhalle," or pump-room, the public gardens, and numerous villas and boarding houses. This part of the town is exclusively populated by the foreign element, and it would be difficult to find a dozen large houses here which are not either hotels or lodging-houses.

The principal spring of Baden is the Ursprung, which

resembles the Kochbrunnen of Wiesbaden, being a warm
saline water, but is much weaker.

Analysis of the Ursprung, by Bunsen.

		Grains.
Chloride of sodium,	. . . .	16·52
Chloride of magnesium,	. . . .	0·09
Chloride of potassium,	. . . .	1·25
Phosphate of lime,	. . . .	0·02
Sulphate of lime,	. . . .	1·55
Sulphate of potash,	. . . .	0·01
Carbonate of lime,	. . . .	0·88
Carbonate of magnesia,	. . . .	0·02
Carbonate of ammonia,	. . . .	0·03
Carbonate of protoxide of iron,	. . .	0·91

Total solid contents in a pint of the water, . 21·35
Carbonic acid gas rather less than a cubic inch.

The Baden-Baden waters are used both internally and
externally. The cases in which Baden is resorted to are very
similar to those in which Wiesbaden is indicated. During
each of my visits to Baden-Baden I met with patients suffering
from a mild form of chronic rheumatism, who seemed to have
derived benefit from this remedy. It is also admissible in
some cases of neuralgia ; and Dr Edwin Lee records its
virtues in instances of "nervous affections of a convulsive
nature, such as hysteria, with congestions of internal organs
and irregularity in the performance of periodical functions."*
It is prescribed with occasional advantage in the treatment of
dyspepsia, and in irritability of the vesical and intestinal
mucous membranes.

Dr Seegen† says that Baden water is indicated in the
treatment of chronic catarrh of the respiratory organs. I
should, however, hesitate to send any patient of mine suffer-

---

* "The Baths of Germany," by Edwin Lee, M.D., 3d edition, p. 168.
† "Handbuch der Heilquellenlehre," Seegen, p. 446.

ing from chronic bronchitis to so variable a climate as this, unless the disease were clearly connected with the gouty diathesis. It is also employed in scrofulous affections of the glands and skin, and, by the resident physicians, in almost every other disease.

CHAPTER XL.

THE SWISS BADEN.

BADEN-ON-THE-LIMMAT, in the Canton of Aargau, is the oldest known watering-place in Switzerland, and may be reached by railway from Bâle in a couple of hours. The town, which is evidently of great antiquity, is situated in part on a kind of platform overhanging the river, and in part in the ravine through which the stream flows. At the south end of the place a curious antique covered bridge crosses the river, and at the opposite extremity are situated the mineral baths and springs. The principal source issues immediately in front of the Schweizer hotel.

This spring is warm, very gaseous, and strongly sulphurous.

Analysis of Baden Mineral Water, by Löwig.

				Grains.
Sulphate of soda,	.	.	.	2·218
Sulphate of magnesia,	.	.	.	2·442
Sulphate of lime,	.	.	.	10·860
Chloride of potassium,	.	.	.	0·711
Chloride of sodium,	.	.	.	13·042
Chloride of magnesium,	.	.	.	0·566
Carbonate of lime,	.	.	.	2·599
Carbonate of magnesia,	.	.	.	0·152
Carbonate of strontium,	.	.	.	0·005
Fluoride of calcium,	.	.	.	0·016
Chloride of calcium,	.	.	.	0·719
All other ingredients,	.	.	.	0·013
Total solid contents,	.	.	.	33·343

Baden was frequented as a watering-place by the Romans,

and numerous relics of this bath-loving people have been
found in the environs of the springs. In the Middle Ages
Baden was more resorted to than at the present day, and
Poggio Bracciolini, the celebrated Roman scholar and courtier
of the fifteenth century, in a letter written in 1415, has left a
graphic account of a fashionable watering-place of that day :
" I write to you," he says, " from these baths, to which I have
now come, to try whether they can remove an eruption which
has taken place between my fingers ; to describe to you the
situation of the place, and the manners of its inhabitants,
together with the customs of the company who resort hither
for the benefit of the waters. . . . They are resorted to by
males and females, who are separated by a partition. In this
partition, however, there are windows, through which they
can converse with each other. Above the baths are a kind of
gallery on which the people stand who wish to see and
converse with the bathers; for every one has free access to the
baths, to see the company, to talk and joke with them. The
bathers frequently give public dinners in the baths, on a table
which floats on the water . . . and they spend the greater
part of their time in the baths, where they amuse themselves
with singing, drinking, and dancing."*

Two centuries later, an old English traveller gives the
following description of the baths of the Swiss Baden, as they
existed in 1617 :—" These baths," says Fynes Moryson, " are
famous for medicine, and are in number thirty, seated on each
side the brooke, which divideth them into *Bethora*, the great
and the little. In the great, divers baths are contained under
one roof of a house, and without the gate are two, common to
the poore. These waters are so strong of brimstone as the
very smoak warmeth them that come neere, and the waters

* "The Life of Poggio Bracciolini," by the Rev. William Carpenter, pp.
69-76, Liverpool, 1802.

burn those that touch them. Of these, one is called the
Marques Bath, and is so hot as it will scald off the haire of a
hogge. The waters are so cleere as a penny may be seen in
the bottome, and because melancholy must be avoided, they
recreate themselves with many sports, while they sit in the
water; namely, at cards, and with casting up and catching
little stones, to which purpose they have a little table swim-
ming upon the water, upon which sometimes they doe likewise
eate. These baths are very good for a cold braine, and a
stomach charged with rhume ; but are hurtful to hot and dry
complexions, and in that respect they are held better for
women than men."*

At the present time Baden-on-the-Limmat is frequented
by few except Swiss valetudinarians, or invalids from the
neighbouring German states, during the season, which lasts
from June to September. It is a quiet, and somewhat
triste, though beautifully-situated watering-place. These
baths and waters are principally employed in the treatment
of some obstinate skin diseases, in secondary and tertiary
syphilis, and also in chronic rheumatism and gout, with
exudation into the joints.

---

* "An Itinerary written by Fynes Moryson, Gent., containing his Ten
Years' Travels," &c., Part 1. bk. i. p. 26, folio, London, 1617.

CHAPTER XLI.

SCHINZNACH AND WILDEGG.

RETRACING our steps towards Bâle, half an hour's journey by train from Aargau brought us to Schinznach where, as at some other Swiss watering-places, there is neither town nor village, but merely a vast bathing establishment, in which those who go through a course of the water must live in community, isolated from the rest of the world.

This spa may now be easily reached from London, either by Paris and Strasbourg, or by the Great Luxemburg Railroad, via Bâle, in two days. The bath-house, which overlooks the river Aar, was commenced in 1695 and reconstructed recently, contains excellent accommodation for about four hundred visitors. At a short distance from this is a sanatorium, where seventy-six poor patients are well cared for at a nominal charge.

Schinznach is the strongest sulphurous water in Switzerland. The following is a recent analysis of this source:—

		Grains.
Chloride of sodium,	. .	5·001
Chloride of potassium, } Chloride of ammonia, }	.	0·063
Sulphate of soda,	. . . .	0·919
Sulphate of lime,	. . . .	4·886
Sulphate of magnesia,	. . . .	2·052
Carbonate of lime,	. . . .	1·086
Carbonate of magnesia,	. . . .	0·063
Aluminia,	. . . . .	0·045
Silicic acid,	. . . . .	0·086
Total solid ingredients,	. .	14·201

			Cubic Inches.
Sulphuretted hydrogen gas,	.	.	1·268
Carbonic acid gas,	.	. .	1·886
Nitrogen,	.	. . .	Traces.

The most important constituent in this water is the sulphuretted hydrogen gas, the physiological effects of which vary according to the dose,—a small dose being a tonic, while a large one is a very powerful stimulant. Its principal use is in the baths. When used internally, it must first be allowed to remain exposed to the air for some time after being drawn from the source, for the purpose of admitting of the escape of a large proportion of the gas it contains.

When drank with the precaution I have mentioned, the Schinznach water acts as a stimulant and resolvent; it excites the activity of the gastro-intestinal mucous membrane, accelerates the pulse, and determines to the skin. According to Dr Amsler, this spa has a peculiarly stimulating effect on the pulmonary mucous membrane, being specially contraindicated and likely to be injurious in any bronchial or pulmonary complaint, and in any organic visceral disease whatever, or even for patients of a full plethoric habit of body.

A course of the Schinznach baths generally produces a specific cutaneous eruption, which is regarded as a proof that the patient's system is under the influence of the remedy. This point is insisted on as of great importance, and is watched for with the same care that we examine the gums of a patient under a mercurial course.

Dr Amsler and other local writers believe that nearly all chronic skin diseases, more especially eczema, psoriasis, and pityriasis may be cured by these baths.

Scrofula, next to skin diseases, is the malady which brings most patients to Schinznach. The stimulant action of the water on the glandular system renders it peculiarly adapted

P

for scrofulous diseases, either external or mesenteric. In cases of chronic rheumatism I was here assured that this spa is almost a specific, and in some cases of dyspepsia and in diseases occasioned by deficient or irregular menstruation, the stimulant properties of this very powerful sulphurous spring are occasionally beneficial, but in all cases, for reasons already stated, it must be used with much caution and under medical advice.

The season lasts from May to September, inclusively, and the mode of using the water is the same as at other strong sulphurous baths, externally in douche, vapour, and other baths, as well as internally in doses of from one to two small glasses, morning and evening.

Close to Schinznach is Wildegg, where a strongly saline iodated spring was discovered in 1830, the water of which is now largely exported.

The source of Wildegg rises through an artesian well some three hundred feet deep, which furnishes so small a supply that it can hardly fill fifty small bottles daily. Wildegg belongs to a class of mineral waters of which we have comparatively few examples, namely, the iodated and bromated spas. According to Dr Laué,* the following is the analysis of this spring :—

				Grains.
Iodide of sodium,	.	.	.	0·218112
Bromide of sodium,	.	.	.	0·236544
Chloride of sodium,	.	.	.	80·236800
Chloride of potassium,		.	.	0·039936
Chloride of lime,	.	.	.	1·980672
Chloride of magnesium,		.	.	12·451584
Chloride of strontium,		.	.	0·152832
Hydro-chloride of amonium,		.	.	0·049152
Sulphate of lime,	.	.	.	14·172672

* " Etudes sur les Eaux Minerales de Schinznach et de Wildegg," par A. Hermman, p. 22.

Analysis of Wildegg continued.

		Grains.
Nitrate of soda,	. . .	0·339456
Carbonate of lime,	. . .	0·583680
Carbonate of iron,	. . .	0·061400
Silicic acid,	. . . .	0·030720

Total solid contents of 16 ounces,	.	110·553600
Carbonic acid gas, . .		2·36 cubic inches.

Small as the amount of iodine and bromine in the Wildegg spa may seem, when compared to the great strength of its other saline constituents, it is to these salts that it owes its reputation as a remedy of remarkable efficacy in the treatment of chronic glandular and scrofulous diseases; the iodine and bromine stimulating the absorbent system, and thus dissipating indolent tumours.

CHAPTER XLII.

RAGATZ, PFEFFERS, THE ENGADINE, ST MORITZ, TARASP, AND LEUK.

COMPARATIVELY few travellers for health now resort to Pfeffers or Pfäffers, as some call it, although many visit the adjoining village of Ragatz, which has become one of the most frequented watering-places in Switzerland, being within nine hours' journey of Zurich by railway, and three-quarters of an hour from Coire on the road to the Via-Mala. The situation of Ragatz, at the entrance of the wild gorge of the Tamina, and the view from the hotel of the extensive valley surrounded with snow-clad mountains which extend before it, are most picturesque. The principal hotels, the Ragatz-Hoff and Hotel de Tamina, are excellent, and have commodious bathing establishments attached to them. But as these are supplied from Pfeffers, some miles distant, and as the curative effects of this water are, probably, more connected with its thermal condition than with its chemical composition, there can be no therapeutic advantage in using the baths at Ragatz instead of at their source.

From Ragatz, a steep, winding road, immediately behind the hotel, brought us to the cliffs which overhang the Tamina, along the left bank of which the path is cut through the rocks. This road here and there descends to the level of the water and then rises to the edge of the precipice above it. In some places it is carried through short tunnels in the rock; and nowhere is the route devoid of picturesque beauty, and often passes through wild and sublime scenery.

The ancient convent, now the bath-house of Pfeffers, is built on a narrow ledge of rock above the torrent, and is over-shadowed by the opposite precipice, which, rising five hundred feet, keeps the house in perpetual shade. The building is a long, narrow edifice, six stories high, and at the time of our visit, when the season was nearly over, looked exactly like a deserted cotton factory. This establishment, which is the property of the government of the canton St Gall, contains accommodation for about three hundred visitors.

Immediately beyond the bath-house the ravine of the Tamina contracts to a crevasse not quite thirty feet in width, over which the precipices on either side, sloping towards each other, form a limestone roof four hundred feet high, and thus enclose a long cavern lighted by the few rays that find their way through the fissures where the rocks above meet. Along this the pathway to the thermal sources passes for nearly half a mile, midway between the roof and the abyss through which the torrent rushes; the passage being supported partly on a narrow ledge of rock projecting over the river, and in part on stakes driven into the marble walls of the crevasse.

At the extremity of this pathway are situated the thermal sources to which Pfeffers owes its fame. They both rise within a few yards of each other, at the bottom of a cavern on the right bank of the Tamina. These springs have neither taste nor odour, are perfectly limpid, have the specific gravity of common water, and issue from the fountain at the temperature of 99°.

The thermal source of Pfeffers was first discovered in 1038, by a huntsman who, alarmed by seeing steam rising from the ground before him, turned back from the chase, and communicated his discovery to the monks of the adjacent convent. By them it was utilised for the benefit of the poor of the district, and its fame gradually spreading, the first thermal establishment was opened in 1242, and patients were admitted—the poor gratuitously, and the rich for whatever offerings they might choose to make on their departure. Thus these baths flourished till 1838, when the government of the canton dispossessed the monks, and turned the monastery into an hydropathic establishment.

From that time Pfeffers has become, year by year, less frequented by invalids, who now seem to prefer the modern establishment at Ragatz. The waters of Pfeffers belong to the same class as Gastein and Wildbad, which possess hardly any chemical ingredients, and depend for their action on their temperature.

Analysis of Pfeffers Spa, according to Pagenstecher.

				Grains.
Sulphate of soda,	.	`	.	.	0·242
Sulphate of potash,	`	.	`	`	0·004
Chloride of sodium,	.	.	.	.	0·208
Sulphate of lime,	.	.	.	0·027
Carbonate of lime,	.	.	.	0·910
Carbonate of magnesia,	.	.	.	0·147
Other salts,	.	.	.	.	0·148
Total solid contents,	`	`	`	1·792

The principal use of the thermal sources ef Pfeffers is in the baths, in which the patients sometimes remain for a considerable time, though now they no longer stay as formerly, when, as an old author assures us,—"Multa dies noctesque thermis non egrediuntur; sed cibum simul et somnium in his capiunt."

The average duration of each bath at present is reduced to about twenty minutes. The same peculiar sense of *bien-être* is ascribed to them as to the Wildbad baths, and it is certain that they exercise a remarkably sedative, but not depressing, influence on the system. After a few days, a slight febrile reaction comes on, during the course of the baths, which is regarded as critical.

Long, indeed, would be the list, if I merely enumerated the diseases which the special writers on Pfeffers and Ragatz say may be cured by these waters. I shall, however, only allude to those ailments which my own experience leads me to think are most susceptible to their therapeutic influence.

Foremost amongst these complaints is dyspepsia and gastralgia, and it is remarkable in what large doses even the most irritable stomach tolerates the water, and how rapidly the best effects—diminution of pain, regular alvine action, and increased appetite—often follow its use.

Nervous and spasmodic affections are frequently benefited by Pfeffers spa, which soothes and tranquillises the nervous system in a special manner, and cases of intractable neuralgia, sciatica, nervous headache, and similar complaints, are sometimes cured by a few weeks' use of these baths, and the internal administration of the waters.

The same remarks may be made of hysteria, chorea, and some other obscure nervous maladies. Also in renal and vesical complaints, such, for instance, as catarrh of the bladder; even when the secretion is attended with pain, the passage through the system of so large a quantity of bland fluid as is daily drank at Pfeffers or Ragatz is likely to be attended with the best effects.

From Ragatz an hour's journey by train brings us to Coire, the ancient capital of the Grisons, where the visitor to the health-resorts of the Engadine must revert to the primitive diligence, by which, in from ten to twelve hours, he may traverse the forty miles of mountain drive by the Julier Pass, which lies between Coire and ST MORITZ. This watering-place, although strongly recommended by Paracelsus in the sixteenth century, was, until recently, almost unknown to British valetudinarians, by whom it is now crowded, not only on account of its chalybeate spa, but still more on account of its pure bracing mountain atmosphere.

The situation of ST MORITZ in the exquisitely beautiful valley of the Upper Engadine, surrounded by the Grison mountains, and immediately below the Julier Pass, at an elevation of six thousand feet above the sea, and overlooking the valley of the Inn and its many lakes, which are generally frozen over until late in summer, although most picturesque, in my opinion renders this village quite unsuitable for the consumptive and bronchitic patients by whom it is now so largely frequented. The climate of the Upper Engadine, bracing and invigorating as it unquestionably is for those who can withstand its low temperature, being not only extremely cold at all seasons, but also probably the dryest in Europe, is far too stimulating for those suffering from pulmonary or bronchial irritation, and should be carefully restricted to those cases which I have pointed out in the first chapter of this work as requiring the tonic influence of a very pure, dry, cold, and bracing atmosphere. To such patients, and especially those suffering from anæmic disorders resulting from over-work and nervous exhaustion, the Upper Engadine, from Maloja to Samaden, may be regarded as the type of that "Happy Valley, wide and fruitful, surrounded on every side by lofty mountains," where Rasselas discovered the tedium of uninterrupted repose, and experienced the irresistible craving "to see the miseries of the world, since the sight of them is necessary to happiness." In truth, there is nothing so remarkable throughout the Engadine as the peculiar stillness which meets the valetudinarian pilgrim in pursuit of health, who may here seek a brief respite from the cares and turmoil of civic life. This stillness, though more broken in the immediate vicinity of St Moritz by the stream of tourists who crowd thither each summer, is still even there sufficiently marked to contrast with any other spa on the continent.

The village of St Moritz, consisting chiefly of half a dozen hotels and a score or so of lodging houses, needs no description. About a mile from the town are the mineral springs to which it owes its repute. The bath-establishment, or Kurhaus, is one of the largest in Switzerland, and is provided with all the usual resources of such places. The springs are powerful alkaline chalybeates, and, according to Dr Planta's analysis, the Alte-Quelle contains eleven grains of saline ingredients in a pint; whilst the Neu-Quelle contains thirteen grains, principally consisting of carbonate of lime, soda, magnesia and iron, and sulphate of soda, together with about forty cubic inches of carbonic acid gas. The class of cases in which a water of this kind should be used have been so fully pointed out in the chapters on other saline chalybeates, such as Spa and Schwalbad, that it is unnecessary to say more than, in addition to the ordinary anæmic, and chlorotic complaints in which they may be prescribed, ST MORITZ, on account of its climatic advantages, may be employed in cases of debility, nervous exhaustion, and premature decay, from over brain-work, pretubercular cachexia, especially in patients of scrofulous diathesis, and in other atonic non-inflammatory chronic disorders.

Tarasp, in the Lower Engadine, may be reached by diligence in eleven hours from the Spa just described. The mineral sources of this comparatively modern, but now very important watering-place, approach in their composition somewhat to the springs we have spoken of in the chapter on the Bohemian "Bitter-wassers," than which, however, they are less saline or purgative, and at the same time are decidedly alkaline and chalybeate, and therefore more alterative and tonic in their action. According to Dr Von Planta, each pint of the Groserquelle of Tarasp contains 39 grains of chloride of soda, 33 grains of carbonate of soda, 22 grains of sulphate of soda, 16 grains of carbonate of lime, 7 grains of carbonate of magnesia, and about a third of a grain of carbonate of the protoxide of iron, together with 32 cubic inches of carbonic acid gas. The principal use made of the saline alkaline waters of Tarasp is in the treatment of chronic abdominal visceral congestions and enlargements, especially of the spleen and liver, more particularly in obstructions of the portal circulation, suppressed hæmorrhoids, intestinal worms, especially tænia, and also, according to Dr Constantine James,* in asthmatic affections.

To the foregoing sketch of the principal Swiss watering-places, a few words must be added concerning LEUKERBAD, or Loëche-les-Bains in the Canton Valais. The most direct route to Leuk is that by railway *via* Berne as far as Thun, and thence by road through the Bernese Oberland, and across the Gemmi Pass. The road from Thun to Leuk, although traversing the most magnificent and wildest of all the Alpine passes, the Gemmi, can hardly be recommended to any nervous valetudinarian, being in many places a mere ledge cut out of the face of the vertical rock which towers above, and from which the nervous traveller can scarcely look without some fear into an abyss of nearly 2000 feet, which lies immediately beneath the narrow shelf on which he stands. Leuk may, however, be now reached by an easier though less picturesque route, namely, *via* Geneva and Lausanne by railway to Sierre, from which station Leukerbad is only five or six hours drive.

The position of this village of hotels, built nearly 5000 feet above the sea at the extremity of an Alpine *cul de sac*, the valley of the Dala, surrounded on all sides by the almost vertical walls of rocky mountains which tower above it, and only protected by a strong dyke of masonry from the fall of

---

* Dr Constantin James, " Guide Practique, Aux Eaux Minerales," p. 457.

some avalanche, such as have thrice already destroyed this watering-place, renders the climate so intensely cold as to be absolutely uninhabitable, except from the end of May until September. The hotels, of which are eight or ten, are, however, crowded by French and Swiss invalid visitors during the season.

The springs of Leuk, some twenty in number, are thermal saline waters, the principal mineral constituent of which is sulphate of lime. The chief of these, the Lorenzquelle, has a temperature of 124°; and the others vary from that to 90°. According to Pagenstcher the following is the

Analysis of the Hauptquelle.

	Grains.
Sulphate of lime, . . .	11·34
Sulphate of magnesia, . .	1·76
Sulphate of soda, . . .	0·45
Sulphate of strontia, ...	0·02
Carbonate of lime, ...	0·31
Chloride of sodium, ...	0·04
Carbonate of potash, ...	0·03
Carbonate of magnesia, ...	0·02
Carbonate of protoxide of iron, . .	0 24
	14·66

The principal use made of the thermal sources of Leuk is in the baths, which are here taken in common, and in which the patients, in full *custume de bain*, of course, remain for the greater part of the day amusing themselves exactly in the fashion which I have in my account of the Swiss Baden quoted from a writer of the sixteenth century. This prolonged immersion in the thermal mineral fluid is unquestionably capable of producing a powerful therapeutic effect in many chronic diseases, and is especially applicable in obstinate cutaneous affections, secondary and tertiary syphilis, and in chronic rheumatism or rheumatic gout, although in the latter complaints the coldness of the climate oftentimes undoes the good effected by the waters.

CHAPTER XLIII.

AIX-LES-BAINS.

OF the French spas frequented by English valetudinarians the most attractive in situation is Aix-les-Bains, close to the Lake of Bourget, ten miles from Chambery, and within fifteen hours of Paris by railway *via* Macon. During the season from May until September, the population of this little town of 4000 inhabitants, is annually increased by some 10,000 invalid visitors, and every available house is then converted into an hotel or lodging-house, yet so great is the demand that it oftentimes is, as I found, a matter of difficulty to obtain any accommodation whatever.

The main feature of Aix-les-Bains is its thermal establishment, a handsome granite building on the hill, a little above the town, and supplied by the two mineral springs which have their sources within the mountains above it. The number of baths contained in the establishment is over three hundred, and includes every variety, from the simple reclining bath to the most complicated local douche, or "pulverised water bath." The *piscinæ* and *vaporarium* are especially deserving of notice. Nearly opposite, and behind the "Pension Chabet," are the ruins of the ancient Roman baths, and the houses in this part of the town are for the most part built from the Roman remains which abound on every side, and some of which—for instance, the handsome triumphal arch of Campanus—still exist in perfect preservation.

Foremost amongst the lions of Aix are the remarkable caves in the mountain behind the town. These, some years ago, were drained by the Government, and thus the supply of thermal water was greatly increased and rendered regular. They are entered by a narrow tunnel cut through the solid rock leading into a lofty circular chamber, with a vaulted dome, which the torches rendered visible. Thence we came into a series of other caverns, some similar to the last, others like vast Gothic churches, with pointed roofs, supported on limestone columns, and many others, in which the limestone, eaten away by the action of the hot sulphurous water, had assumed all kinds of fantastic shapes and resemblances. The exploration into the remoter caverns would be no easy task for an invalid or a lady, and should not be attempted by either.

The springs of Aix-les-Bains belong to the class of warm sulphurous waters. There are two sources—one the sulphurous, the other miscalled the *Source d'Alun*, which, however, contain hardly any trace of alum, and is properly designated the *Source de St Paul*.

Analysis of the Aix-les-Bains Sources, according to Dr Seegen.

A Litre of Water contains (in French Grammes)	Sulphurous Source.	Alum Source.
	Gramme.	Gramme.
Carbonate of lime, . . .	1·1803	1·2384
Carbonate of iron, . . .	0·0387	0·0774
Chloride of calcium, . . .	...	0·4644
Chloride of magnesium, . . .	0·1548	0·1548
Sulphate of lime, ...	0·4257	0·6966
Sulphate of magnesia, . . .	0·7353	0·2322
Sulphate of soda, . . .	0·3483	0·2322
Baregine or glairine, . . .	A trace.	A trace.
Total, ...	2·8831	3·0960
Temperature, . . .	115° (Fahr.)	117°

The alum water is that generally taken internally, though comparatively little internal use is made of the waters at Aix.

It is in the baths, and especially in the douche baths, that their efficacy is most often proved.

The diseases in which a visit to Aix are commonly recommended are—chronic rheumatism, especially when enlarging and disabling the joints ; rheumatic gout; and above all in certain chronic skin diseases. The local douches are also applied with great benefit to diseases of the eye and ear, and the so-called " pulverised water," is used with advantage in cases of clergyman's sore throat and in ozœna.

Three-quarters of a mile from Aix-les-Bains, on the road to Chambery, is the spa of MARLIOZ, where a *sale de inhalation*, and pump-room, has been constructed within the last few years. This building contains a couple of rooms—one for gentlemen, the other for ladies, where the water is forced up and reduced into a fine spray, which is inhaled for various periods by the patients.

Marlioz is a cold, strongly sulphurous spring, containing traces of iodine. Its action is stimulant, and it is employed internally, and also in local douches, as well as by inhalation. By the physicians of Aix it seems regarded as almost a specific in pulmonary diseases, especially chronic bronchitis.

CHAPTER XLIV.

VICHY AND THE MINERAL SPRINGS OF AUVERGNE.

VICHY, which is the spa *par excellence* of France, is situated on the Allier, in the department of the same name, and within ten hours' drive of Paris, by the Lyons Bourbonnais railway.

The town has a resident. population of about 6000 inhabitants, and is divided into two distinct parts, viz., *Vichy les Bains* and *Vichy la Ville*, which are separated by the park. The former consists of two or three long, handsome streets of hotels and lodging-houses, and here almost all the visitors reside. Neither the modern nor the older part of Vichy present much deserving of special notice, excepting the mineral springs, the new Casino, and the thermal establishment.

The environs are said by several writers to be flat and uninteresting, but this observation, which is evidently copied from a well-known guide-book, does not apply, except to some parts of the immediate vicinity of the town ; for the neighbouring country, at a short distance from Vichy, presents some of the most beautiful scenery and most interesting excursions in France.

The geological strata from which the springs arise are the tertiary limestone and coal formations, and under the rock the basin of mineral water is supposed to extend over an area of six miles.

There are nine mineral springs in use in Vichy, which are alkaline, ferruginous, and highly charged with carbonic acid gas, and in all, minute traces of arsenic have been discovered. They are divided into cold and thermal, also into natural and artesian. The former are in general warmer and less gaseous than the latter, which contain most iron. In all the principal saline ingredient is bicarbonate of soda.

Analysis of the principal Mineral Sources of Vichy (M. Mossier).

Substances in One Pint of Water.	Grande-Grille (grains).	Puits Carré.	Petit Puits Carré.	Le Hôpital.	Boulet.	Célestins.	Lucas.
Carbonate of lime, .	1·61	1·70	2·30	2·45	3·29	Analysis imper- fect.	3·41
Carbonate of mag- nesia, . .	0·30	0·30	0·30	0·27	0·35		0·41
Carbonate of iron, .	0·08	0·15	...	0·36	0·35		0·17
Carbonate of soda, .	34·61	32·40	36·30	33·52	42·70	32·07	28·56
Sulphate of soda, .	5·57	6·91	7·05	6·27	3·04	5·46	6·49
Chloride of sodium,	3·15	3·88	2·64	1·10	0·47	3·45	7·31
Total, . .	52·13	45·34	48·59	43·37	50·20	40·98	46· 5

Under the gallery behind the thermal establishment are found four of the mineral springs. The principal of these is the Grande-Grille, so named from the iron railing that formerly surrounded it, the temperature of which is 108°. Its taste is rather disagreeable—saline, and somewhat ferruginous. It is employed internally and externally, and, as it keeps well, is largely exported. This source is principally used in gout, gastric complaints, dyspepsia, and affections of the liver.

The "Source des Mesdames" is non-thermal, and is nearly identical in composition with the "Puits Lardy," containing a large proportion of salts of iron, with traces of arsenic. As its name imports, it is used in diseases peculiar to women, and especially in anæmia and chlorosis.

The "Celestins" and the "Puits Lardy" are situated at the extremity of the old town, on the right bank of the river. The latter is the only one of the sources which does not belong to the company which now farms the wells of Vichy. They contain more carbonic acid than the other springs. According to Dr Barthez,* the most eminent authority on the subject, the "Puits Lardy" contain also a large proportion of sulphuretted hydrogen gas, and being, therefore, the most stimulant of the Vichy waters, cannot be used with safety by those suffering from diseases attended by inflammatory or hæmorrhagic symptoms, being principally employed in renal or vesical complaints, and in anæmic cases.

The Vichy waters are alkaline, aperient, and alterative. Their principal use is in the treatment of gout, and in chronic diseases of the stomach, or abdominal viscera, such as dyspepsia, chronic hepatic disease, biliary calculi, fatty degeneration, or cirrhosis, and in hæmorrhoidal affections, which are so often connected with congestion of the liver. They are equally serviceable in enlargements of the spleen, and in many cases of hypochondriasis. Moreover, this spa is specially adapted for the cure of some of the chronic diseases of women connected with disordered menstruation, and for the anomalous "critical complaints" which often set in at the period of life when this function ceases. It is also much prescribed by French physicians in cases of diabetes, and in some instances may be thus used with great benefit.

The complaint for which nine-tenths of the English visitors drink these springs is gout, and I believe that the disappointment which so often drives gouty patients home again to patience and flannel, is the result of the misconceived ideas which prevail on this subject. It should be distinctly

* "Guide Pratique Aux Eaux de Vichy," par le Dr F. Barthez, 7th ed., p. 94.

Q

understood that Vichy water is not a specific for the gout. It can only act on the gouty diathesis, by improving the tone of the digestive organs, augmenting the secretions, and correcting the abnormally acid condition of the blood in such cases. Gouty patients whose disease springs from dietetic errors and neglect of exercise, come to Vichy ; their appetite is increased by the change of air and foreign cookery ; they indulge that fictitious appetite fully at the table d'hôte, and then return home wondering why they ever went so far for so little good. The remedies for gout are abstemiousness and exercise. Vichy water may aid, and aid materially, but it cannot supersede these.

Two miles from Vichy is the town of CUSSET, the antiquity of which is evinced by its narrow winding streets, quaint old houses, and handsomely-wooded boulevards. Entering it, we passed a large tower, now used as a prison, whose walls, twenty feet thick, resisted many a fierce attack from the lords of Auvergne and Bourbonnais in the troublous days of Louis XI. Now, however, Cusset is an unimportant market town, remarkable only for its mineral waters, which seem to have attracted less attention than they deserve.

These springs belong to the same class as those of Vichy, than which they are, however, stronger, containing more carbonic acid gas, bicarbonate of soda, and iron. The "Elizabeth" well, for instance, is said by its proprietors to contain six times as much soda as any of the Vichy springs. The " St Marie " is also very rich in the same salts. It supplies the bath-house, a handsome building with reading saloons, pump-rooms, and about thirty very neat and well-constructed douche and reclining baths.

The waters of Cusset, which are all cold, are used in the same class of cases as those of Vichy, but being stronger, require still more caution in their administration.

In the same ancient province of Auvergne, though not in the same department, there are some other mineral springs, less known than Vichy to Engish valetudinarians, but esteemed of considerable remedial power by French physicians and their patients. For the materials of the following brief notice of these spas I am indebted to my father, Dr R. R. Madden, who has visited them since I have been in Auvergne.

The spas to which I would now invite attention are the thermal waters of Mont-Dore, St Nectaire, and Royat. The first of these, MONT-DORE, may be reached from Vichy by railway to Clermont, and thence by coach. It lies about thirty miles from Clermont, in a small valley 3500 feet above the sea, and immediately under the Pic du Sancy, the highest mountain in central France. Between Clermont and Mont-Dore the road which passes the remarkable mountain of the Puy de Dôme crosses the most singular, and to a geologist most interesting, volcanic district in Europe; on every side may be seen extinct craters, masses of the scoriæ and lava ejected from these, and vast blocks of basaltic rock, evidently of volcanic nature, all of which attest that this region was at one time the scene of convulsive igneous action of incalculable force and activity.

MONT-DORE-LES-BAINS, although said to be one of the most ancient watering-places in Europe, is now but a village, containing several good hotels. There are eight mineral sources here, the temperature of which vary from 115° to 59°. The chief chemical ingredients in all are bicarbonate of potash, carbonate of lime, and sulphate of soda. Besides these, recent chemists have proved that the waters contain rather more than one millegramme of arsenite of soda in each litre.

The cases in which the baths and waters of Mont-Dore are prescribed are certain forms of chronic bronchitis, asthma, and laryngeal complaints, gastro-enteric, and uterine disorders marked by congestion, similar cases in which the liver is implicated, nervous maladies, such as neuralgia and sciatica, and scrofulous diseases, especially of children.

About fifteen miles from Mont-Dore, near Murol, is the watering-place of ST NECTAIRE, also in a volcanic district. There are seven thermal springs in this locality, the temperature of which varies between 75° and 110°. They are all alkaline, ferruginous, and stimulant. They are principally used in cases of renal and hepatic disease, in enlargements of the liver, or spleen; and are also employed in amenorrhœa, leucorrhœa, and gout.

The last of the spas of central France to which I shall allude is ROYAT, situated eight miles from Clermont. The waters of Royat closely resemble those of Mont-Dore, than which they are, however, one-third stronger. Royat is in considerable repute with many French physicians in the treatment of scrofula, gout, and rheumatism.

CHAPTER XLV.

THE SPAS OF THE PYRENEES—CAUTERETS.

IN no part of Europe will the valetudinarian find so wide a choice of mineral and thermal springs to select from, within the same extent of country, as in the Pyrenees, where some two hundred of these fountains of health have been discovered.

The mineral waters of the Pyrenees may be divided into three classes, viz.:—Sulphurous, Saline, and Ferruginous, and two-thirds of them belong to the first-named class, of which Cauterets, Barèges, Bagnères-de-Luchon, and Saint Sauveur are examples. The saline waters are illustrated by Bagnères-de-Bigorre and Dax; and the ferruginous by Castera-Verduzan and Casteljaloux.

The attractions of the Pyrenees are not, however, confined to the invalid traveller, but even for the pleasure tourist offer inducements for a pedestrian excursion in some respects superior to any in Switzerland. And for a man in health, what mode of travel affords hereafter such pabulum for memory, such a variety of incidents, and such pleasant recollections, as a pedestrian journey with a genial companion, and in fair weather, through so beautiful a country as the High Pyrenees. He who would attempt this, however, must be prepared to "rough it;" to endure fatigue, occasional inclemency of weather, meagre diet, and indifferent lodging, if he would go beyond the mere beaten track of tourists; and to dispense for the time with the luxuries, contenting himself with the necessaries of life.

It is strange how soon one gets accustomed to the hardships of this mode of life. Thus, for instance, although previously unused to pedestrianism, within a few days after I commenced this walking tour, I found myself so braced up by the pure mountain atmosphere, that I could walk without any material inconvenience from early morning until evening.

Such a programme may not look inviting, but this I know, that if the object of an autumn tour abroad be the improvement of health impaired by attention to some absorbing pursuit, and a sedentary civic life, a pedestrian journey of this kind will do the traveller more physical, as well as moral good, than would the same time spent in almost any other way. And even for valetudinarians who are unable or unwilling to encounter hardships and fatigue, the Pyrenees offer resources. The railroad from Bordeaux now runs to the very centre of these mountains, and every spa is within an easy drive from the train.

Our pedestrian journey through the Pyrenean watering-places commenced at Tarbes, ten kilometers from which the ascent of the mountains begins. After a brief pause in the hamlet of Mont-Gaillard, where the hostess of the village inn prepared an omelet worthy of the Maison Dorèe, leaving the tilled plain behind, we entered into a hilly pasture country, not less populous than the lowlands, and at nightfall arrived at Bagnères.

BAGNÈRES-DE-BIGORRE is an ancient town of 9000 inhabitants, situated at the foot of the mountains, between the valleys of Tarbes and Campan ; and may now be reached from Paris by railway, *via* Bordeaux, in thirty-six hours.

The aspect of Bagnères-de-Bigorre is very Spanish. The promenade is more like an Andalusian " Alameda " than a

French Boulevard; and the narrow winding streets, the projecting roofs of the houses, and the dress of the peasantry, all reminded me of a Spanish scene. The bath-houses and thermal establishment are large, handsome buildings, and the hotels are numerous and good.

There are a considerable number of mineral springs in Bagnères-de-Bigorre belonging to the class of saline sulphurous waters; the town being probably built over a subterranean thermal stream, which issues forth wherever an opening is made; and consequently the several wells differ only accidentally, according to the strata they may pass through between this subterraneous river and the surface. Besides the sulphate of lime which is characteristic of all these sources, with few exceptions, they contain more or less carbonate of iron; and act as stimulating sulphurous saline chalybeates. Accordingly they are indicated in the treatment of anæmic and chlorotic cases; in chronic mucous discharges from either the urinary or the pulmonary organs, when unaccompanied by any inflammatory action; in hæmorrhoids; habitual constipation; dyspepsia and loss of appetite, and in some forms of enlargement of the liver and spleen.

The waters of "La Reine," and of "Lasserre," in doses of from five to six glasses, are considered as mildly laxative, as well as stimulant. The less strongly mineralised sources —"Le Source Foulon," "Petit Baréges," "Salut"—are said to produce a somewhat sedative action on the nervous system.

Having remained some days in Bagnères we resumed our journey to Cauterets, visiting on our way Lourdes, now the most famous shrine in Europe. On leaving Lourdes, which we did at a very early hour, we pursued our way by a road not unlike that which those who have crossed the Irish Channel will remember as the Scalp, near Dublin, with mountains on all sides, whose loose rocks seem ready to fall on the traveller below. Emerging from this desolate region we passed through the plain of Argalèz, the loveliest valley of the Pyrenees; and making a brief halt at the village of Pier-refitte, which closes this vale, and where two gorges open in the mountains, one on the left leading to Baréges, and the other on the right to Cauterets, arrived at our destination after a walk of ten hours.

CAUTERETS is situated some 3000 feet above the level of the sea, in a deep *cul de sac*, formed by snow-covered mountains, the only road through which, beyond the village, being by the foot paths which lead into the recesses of the Pyrenees, or into Spain. The houses of Cauterets form a long narrow street divided by an irregular square containing the "Hotel de Paris" and some others, whence a shorter street leads up the side of the mountain to the thermal establishment, which is fitted up with every modern improvement in the baths, and is supplied from the distant mineral sources on the hill, in the same way as the waters of Pfeffers are brought to Ragatz.

The railway *via* Bagnères-de-Bigorre, has now placed this watering-place within a couple of days' journey from Paris. Moreover, the beauty of the surrounding scenery, and the excursions which may be made from hence through the High Pyrenees and into Spain, especially that by the "Pont d'Espagne" to the wild and romantic Lac de Gaube, attract many visitors to Cauterets, who have no need of its mineral springs.

Cauterets possesses twelve thermal sources, which are scattered about the village, and some are at a considerable distance from it. For convenience of description they are divided into two groups, viz.:—*Les Sources de l'Est* and *Les Sources du Midi.* They differ in temperature from 131° to 86° and present some variety of composition, though they all agree in being rich in sulphuretted hydrogen gas, in silica, and in glairine, and in the rapidity with which the sulphates they contain are decomposed and changed into sulphites or hyposulphites.

LA RAILLÈRE, the most celebrated of the mineral springs of Cauterets, is about half an hour's walk from the town, on the road to the Lac de Gaube. The water is clear, and its temperature is 102°; it is saponaceous to the touch, with a slightly sulphurous smell, and sweetish, mawkish taste. It is conducted into a commodious bath-house, the douches in which are particularly well constructed. This spring is used in chronic catarrhal affections of the respiratory organs, and in incipient phthisis. I do not myself, however, agree with those who prescribe this remedy in the latter disease.

The sources of " César " and " Les Espagnols," which supply the grand bath-house, are the most stimulating waters of Cauterets. Internally, they are used in chronic catarrh, and in some forms of asthma, and, as baths, they are ordered in cases of chronic rheumatism, certain skin diseases, and scrofulous affections.

" Les deux Pauces " are similar in action, but milder than the last-described springs. The " Petit St Sauveur " is still more soothing, and is chiefly used for baths and douches, in leucorrhœa and some uterine diseases, as well as in certain nervous affections. " Le-Pré " is administered in chronic rheumatism ; and so also are the sources of "Mahourat " and " Le Duis," which are nearly a mile from the town.

CHAPTER XLVI.

THE PYRENEAN WATERING-PLACES CONTINUED.

FROM Cauterets we returned to Pierrefitte, whence the road to Lux and Barèges branches off to the left, and after a steep ascent of four miles enters the dark and cheerless valley of the Bastan, in which the spa we are about to describe is situated.

BARÈGES is a small village, consisting of one long street, of about a hundred houses, built of stone, standing immediately over a mountain torrent, the Gave of the Bastan, and is as wild and desolate a place as can well be imagined.

Being the most elevated watering-place in Europe, the climate, even in summer, is cold and variable, and in winter it is such as to render the village uninhabitable. None but those who absolutely require the waters are to be met with in Barèges, for nothing else could, I think, induce any one to pass a single week in this village. And yet, however, between 6000 and 7000 invalids reside here during the short season. Great, therefore, must be the medical virtues of the springs, which can thus attract the votaries of fashion and of pleasure to so remote, inaccessible, and dreary an abode as this.

There are nine mineral sources in Barèges, the temperature of which vary from 86° to 112°. Eight of these are contained within the bathing establishment, in the centre of the town. They are perfectly clear, have a strong sulphurous taste, and well-marked "hepatic" odour. They are more fixed, and change less by keeping than the other Pyrenean sulphurous waters, and they all contain a considerable amount of the peculiar, pseudo-organic, unctuous substance called "Barègine."

The saline matters found in the nine sources of Barèges principally consist of the sulphates of soda and lime, silicates of the same salts, chloride of sodium, with traces of oxide of iron and iodine.*

The primary action of the Barèges spa is stimulant and tonic, producing considerable nervous and vascular excitement; accordingly, it is best suited for persons of lymphatic or scrofulous diathesis, and it should be especially avoided by those of a plethoric habit of body, by pulmonary invalids, as well as by valetudinarians suffering from any hæmorrhagic or congestive disease.

In the form of baths, these waters are applicable in the treatment of cases of old wounds, either breaking out afresh or causing recurring pain, in carious disease of the bones, in articular complaints, whether rheumatic or resulting from injury. They are largely used in obstinate skin diseases, more especially when of secondary, or of mercurial origin; also in chronic rheumatism and scrofulous affections. The duration of the course must be regulated by the effect it produces. The full time is six weeks; but in many cases half that period might be dangerous to the patient's life.

About a mile from Lux is a spa in every respect different from that which has been last described, viz., ST SAUVEUR. The situation of this village, at the entrance of the plain of Gavarnie, within reach of the most sublime scenery in France, i.e., the Brèche de Roland, Mont Perdu, and the Vignemale, is most attractive. The watering-place itself is a mere hamlet connected with Lux by an avenue of poplars. There are a

* MM. Petrequin et Socquet, "Traite Général des Eaux Minérales," p. 413.

couple of good inns, besides which the accommodation of the lodging-houses is comfortable, the charges are moderate, and the thermal establishment is very complete.

There is but one mineral source in St Sauveur, although there are four or five outlets from this, which some writers mistake for distinct springs, and describe as such.

The water is saline and alkaline. It is perfectly limpid, is warm, 95°, and has a soapy or unctuous taste, caused by the amount of glairine suspended in it.

The waters of St Sauveur are used as baths, injections, and for drinking. Their medicinal action seems due to their temperature, the amount of nitrogen and glairine they contain, and their alkalinity. In their influence on disease they somewhat resemble the springs of Wildbad, and exert a peculiarly sedative influence on persons of a nervous and irritable temperament. They are prescribed in hysterical cases, in dyspepsia, in vesical catarrh, in chronic ovarian or uterine affections, and leucorrhœa, and in neuralgia and sciatica.

The next of the Pyrenean spas, in the order of their position, is BAGNÈRES-DE-LUCHON, which lies close to the Spanish frontier, and may now be reached in about forty hours from Paris by the railway to Bagnères-de-Bigorre, and thence by diligence.

The accommodation for invalid visitors at Luchon is better than at most of the Pyrenean spas, there are nine or ten very tolerable hotels, and a fair choice of apartments. Living is comparatively cheap and good, game being abundant, and the fish from the mountain streams is excellent. Not only are the comforts, but the luxuries of civilised life, such as balls, concerts, clubs, &c., attainable in Luchon, and in this respect it comes nearer to the German watering-places than the other Pyrenean health resorts.

There are about thirty distinct mineral springs in Luchon,

the temperature of which varies from 150° to 60°. These are all sulphurous and saline; their principal saline contents being sulphate and hyposulphate of soda, with silicate of soda.

The water is used principally for bathing, but is also taken internally, in doses of from two to three small glasses, either pure or with an equal part of milk. The diseases in which Luchon is resorted to are—cutaneous affections, indolent ulcers, chronic rheumatism, arthritis, and caries. It is also strongly recommended in many cases of scrofulous enlargements, hypochondriasis, and dyspepsia.

AMÉLIE-LES-BAINS, in the department of the Eastern Pyrenees, on the route from Perpignan to Barcelona, is now one of the most rising health resorts in the south of France, being frequented by valetudinarians not only in summer on account of its mineral waters, but also in winter by pulmonary invalids, though in the latter case, in my opinion, often with questionable judgment, on account of the mildness of the climate.

The village, which contains about 800 inhabitants, is built in semi-circular form, at the foot of a hill on the right bank of the river Tech, two miles from Arles. The mineral waters of Amélie issue from numerous sources, which are all warm, sulphurous, and very gaseous. There are three separate thermal establishments, principally resorted to in the treatment of chronic rheumatism, skin diseases, neuralgia and sciatica, in affections of the kidneys and urinary organs, leucorrhœa and irregularities of the catamenia. Amélie is also frequented by scrofulous patients, and is said to be productive of great benefit in glandular and articular diseases of this class. It is, moreover, though I believe with much less utility, prescribed in chronic laryngeal and bronchial complaints, and in dyspepsia.

254 THE PYRENEAN WATERING-PLACES.

It would be out of keeping with the plan of this work for me to attempt any detailed account of my walking tour through the remaining Pyrenean spas. But I would however strongly recommend such a tour to the imitation of any traveller for health who has sufficient youth and strength to endure some hardships, such, for instance, as a day's walk from Arruns across the Col de Torte, which our landlord had in vain attempted to dissuade us from crossing, as he said a storm was impending; but mistrusting the disinterestedness of his suggestion, we started. The first persons we encountered after an hour's walk, when we were ascending the mountain, were a couple of herds driving their sheep into the valley, who repeated the same ill-bodings, and advised us to return to Arruns, as the path over the Col was covered with snow. However, as it looked clear and bright we pursued our way, and only when it was too late to turn back we found our error. For, after a long, toilsome ascent, when we had crossed the first Col, and were passing the narrow footpath which overhangs the deep ravine under the Pic de Gabiscos, a hurricane suddenly arose, drifting snow and sleet, and large stones, were hurled along as dust; we were forced back by the irresistible power of the wind; it was even impossible for us to stand, and it was only by throwing ourselves prostrate that we escaped being blown over the precipice. The wind came rushing down in squalls every few minutes, and between each of these violent gusts we had only time to advance a few perches before we were again forced down. At length, when nearly exhausted, we fortunately reached a wooden shed where we sheltered until the violence of the storm had abated, and crossing the Col de Torte, over which the path was undistinguishable under the accumulated snow, finally, after a walk of twelve hours, arrived at Eaux-Bonnes.

EAUX BONNES is a small village in the Basses-Pyrénées in

the valley of Ossau, about twenty-five miles from Pau. Its situation, in a deep ravine enclosed by lofty mountains, is highly picturesque. These mountains approach so close to the village that they have been quarried in every direction to make way for the modern extension of the place.

The three principal mineral sources, which are found at the foot of Mount Trésor, are "La Vieille" (ou la Buvette), the temperature of which is 92°; "La Nouvelle," temperature 88°; and "La Source d'Eau bas," temperature 90°. Besides these there are two other springs immediately outside the village; one of these wells, which is cold, is the only source here used internally.

The Eaux Bonnes are all very gaseous, and have a sulphurous smell and unctuous taste. They belong to the class of alkaline sulphurous spas; but are not so alkaline, and contain less silica and more sulphate of lime than other Pyrenean waters of the same class.

In their action Eaux Bonnes are stimulant, though less exciting than some of the Pyrenean sulphurous spas. They require, however, to be used with great caution in small doses, commencing with a quarter of a tumblerful, which may be gradually increased, if it produces no ill symptoms, to two small glassfuls a day.

The season lasts from May to the beginning of September.

Formerly the principal use of the waters of Bonnes was externally in baths, in the treatment of old and painful wounds, and ulcers. Thus their ancient name of "Arquebusades," was derived from the wounded musketeers, who, after the battle of Pavia, were brought hither by Jean d'Albret, to be cured. Now, however, the chief employment of these springs is internally in the treatment of chronic pulmonary affections, among which is included phthisis. My own opinion of the inefficacy of mineral waters in the treat-

ment of consumption has been expressed so often in the preceding pages that I need not here repeat it.

The Eaux Bonnes are most useful in chronic maladies
of the abdominal viscera, in hypochondriasis and hysteria, in
obstinate intermittent fevers, in chronic catarrhal affections,
and, as baths, in the treatment of chronic ulcers, fistula, and
caries.

EAUX-CHAUDES, in the Low Pyrenees, is situated in a wild
and sombre mountain gorge on the right bank of the Gave,
four miles from Eaux Bonnes.

The approach to Eaux-Chaudes, through this dark ravine,
is calculated to depress a nervous patient, and the aspect
of the village itself, consisting of a few straggling houses,
which stand on a narrow ledge between the mountains
and the Gave, is far from cheerful. But still, though presenting nothing but its waters to attract visitors, Eaux Chaudes is
frequented by a considerable number of invalid residents each
season. There are four or five good hotels, and the expenses
of living are moderate.

There are six mineral springs at this spa, all of which issue
at the junction of the granite and limestone formations at
the foot of the mountain, which divides the valley of Eaux
Bonnes from that of Eaux-Chaudes.

The thermal establishment is supplied by three sources—
which are all used both internally and as baths. Notwithstanding its name, none of the springs of Eaux-Chaudes have
an elevated temperature. Thus, the temperature of " Le
Clôt," the warmest of these, is 97°; and one, namely, the
Source Mainvielle, is quite cold. They are all but slightly
mineralised, though rich in " glairine " and sulphuretted
hydrogen gas. The principal saline contents of the waters
are chloride of sodium, sulphate of lime, silicate of lime,
sulphate of soda, and carbonate of soda.

The Eaux Chaudes are very stimulating, acting with equal
energy on the nervous and circulating systems; and are
principally used in the treatment of obstinate skin diseases,
scrofulous swellings, and articular enlargements, in chlorosis
and amenorrhœa, as well as in chronic rheumatism, sciatica,
and neuralgia.

All the Eaux-Chaudes, especially the warmer and more
stimulating sources, require the utmost caution in their use,
and none of them can be employed in safety excepting in
accordance with the advice of a resident physician.

Here ended my walking tour through the Pyrenees. From
Laruns I drove to Pau, and renewed my acquaintance with
the ancient capital of Bearn.

The mineral spring of Pau is situated immediately without
the town, between the park and the river. This source has not
been many years in use, possesses a slight chalybeate taste,
is cold and limpid, and leaves a ferruginous tinge in the
stone basin into which it is received. Its use, which is very
limited, need not be here dwelt on, as it differs in no wise
from what I have already spoken of, in the introduction, as
the general action of all simple chalybeate mineral waters.

Having rested ourselves after our long pedestrian journey
through the mountains, by a short stay in Pau, we turned
our thoughts homewards, and leaving Pau by train in the
afternoon, arrived the same evening in Dax.

R

CHAPTER XLVII.

DAX, PASSY, AUTEUIL, ENGHIEN, PLOMBIÈRES, AND CONTREXVILLE.

THE very ancient town of DAX on the Adour, since the com-
pletion of the railway from Bordeaux to Pau, is seldom visited
by tourists. So far back as the tenth century, the fountain of
" Nelse " was frequented by patients from every part of
Europe ; but now, excepting the inhabitants of the department, hardly a single invalid is attracted to Dax by its
thermal springs. The most important of these is contained
within a large roofless structure of considerable antiquity,
which encloses a reservoir of hot mineral water some seventy
feet in length by fifty in width. The water is clear, and its
temperature in the basin is 125°, whilst in the spring which
supplies this, it rises to 156°.

The scene around the front of the basin when we first
visited it in the early morning was very curious. The whole
population of Dax were apparently assembled in the little
square—every man, woman, and child with a large, peculiarly-
shaped tin vessel, strapped over their shoulders, and each
patiently awaited their turn to fill these at one of the spouts
by which the warm water issues from the basin. I remarked
that before doing so, however, it seemed the rule to take a
deep draught of the thermal fluid. This supplies most of
the domestic and culinary purposes of the people of Dax, to
whom it saves no small expense for fuel.

Besides this there are two other warm springs —" La

Source des Fossés," and "Les Bagnots;" used extensively for bathing.

The mineral springs of Dax, which are almost unknown in this country, are mild, saline, thermal waters, their chief chemical ingredients being the sulphates of lime and soda, with a little common salt, and carbonate of magnesia. They are commonly used, with the happiest result, by the inhabitants of the Landes, in cases of chronic rheumatism, and rheumatic-arthritis, causing impairment of the joints; and in contraction of the muscles, following recovery from severe surgical disease. They were prescribed by the French physicians of the last century, in certain forms of paralysis, and also in pulmonary affections, especially asthma, but are no longer employed in these cases.

Within the recently extended limits of Paris, at PASSY, there exists a very strong chalybeate water—so strong that before it is used internally, it is necessary to allow it to stand exposed to the air for some time, until it deposits a ferruginous sediment which falls rapidly, being only suspended, not dissolved, owing to the want of sufficient carbonic acid gas. There are four sources, whose chemical composition is nearly identical. The chief mineral constituents of all these are, sulphates of iron, lime, magnesia, and soda. The Passy waters are used externally in the treatment of chronic ulcers, and in cases of leucorrhœa. They are also prescribed internally, with the precaution of allowing the water to deposit the suspended salts as I have just described, in doses of from one to three small glassfuls, in cases of general and local anæmia, chlorosis, fluor albus, some intermittent febrile disorders, atonic dyspepsia, and diarrhœa, and, in a word, in all those diseases connected with poverty of blood, in which ferruginous tonics of this class are indicated.

In the neighbouring suburb of AUTEUIL a somewhat similar

mineral water exists. This spring is known as the "source de Quicherat," and from the sixteenth century to the present time has been resorted to by the Parisian bourgeois in the same class of cases as the wells of Passy are employed in.

The last of the watering-places in the neighbourhood of Paris which I visited was ENGHIEN-LES-BAINS, which is within twenty-five minutes' drive of the city by either the Nord or Ouest railways. This spa overlooks the valley of Montmorency and is prettily situated on a small lake surrounded with fantastic villas, many of which are said to belong to persons of the Parisian *demi monde*.

Enghien possesses a large thermal establishment, open all the year round. This building faces the lake, and contains about a hundred bath-rooms, besides excellent accommodation for invalid boarders.

The mineral springs of Enghien are cold, sulphurous waters, containing a large volume of sulphuretted hydrogen gas, and a considerable amount of sulphate of lime.

Of these sources only two, viz., "Du Roi" and "Deyeux," are used for drinking, the others being too strong for that purpose. The Enghien waters are powerful stimulants. In small doses, conjoined with the baths, they increase the appetite, quicken the pulse, and excite a determination of blood to the skin. If taken for some days without intermission, or if the dose be at all too large, they occasion "spa fever," sometimes of a most dangerous type.

It appeared to me, when at Enghien, that these springs are used in a very incautious manner; and in every class of chronic disease without discrimination. I have already shown that strong mineral waters, like all remedies of any value, are two-edged weapons, not less powerful for evil than for good, and those of Enghien are certainly no exception to this rule, and should never be drunk, as they appear to be by some, without judicious medical advice, nor in larger doses than half a tumblerful each morning, nor should the patient take the baths every day, or remain many minutes in them. Both the baths and the internal use of the waters should be discontinued if they produce the least febrile irritation, headache, or cerebral excitement; and the patient should not allow himself to be influenced by the advice of the managers of the thermal establishment, if it be in opposition to these plain rules.

The Enghien spa, like other sulphurous waters, is employed, and sometimes very efficaciously, in the treatment of various skin diseases, scrofulous affections, certain mucous discharges, chronic glandular and articular enlargements, in some "secondary and tertiary symptoms," as well as in rheumatic affections.

A few words on two watering-places in the Vosges Mountains will conclude my account of the French spas. The first of these is PLOMBIÈRES, which is twenty-five miles from Epinal, and may be reached in thirteen hours from Paris by the Chemin de Fer de l'Est. The situation of this town, in a deep ravine surrounded by lofty mountains, renders the climate cold and variable. But despite this, these mineral waters, having been used by Napoleon the Third, are particularly in vogue with the adherents of the late regimé. The place itself is one of the most frequented of the French health resorts, having excellent hotel accommodation, as well as a magnificent bathing establishment.

There are upwards of twenty mineral sources in Plombières; of these fifteen are thermal. Of the non-thermal springs one is chalybeate, and two are what French writers term "sources savonneuses." The general character of all these is a very weak alkaline mineralisation, and their chief employment is in douche and other baths; only two, viz., "La source du Crucifix" and "La source des Dames," being used internally.

The waters of Plombières, as Dr Edwin Lee pointed out
many years ago, in their chemical properties and therapeutic
effect, resemble those of Teplitz, and are not only strongly
diuretic in their effects, but also act as special stimulants on
the uterine system; and hence are prescribed in cases in
which it is desirable to increase the renal secretion, as well as
in the treatment of chronic affections of the uterus and its
appendages, dysmenorrhœa and uterine catarrh.

In the same department as the last place is another locality
which, though long known to contain powerful mineral springs,
has only comparatively recently become a fashionable watering-place, and seems to be in especial favour with American
valetudinarians.   CONTREXEVILLE may be reached in about
twelve hours by the eastern railway from Paris to Neufchateau,
and thence by coach in three hours.

There are three distinct mineral sources in Contrexeville, the
most important of which is " La Source du Pavillon," but all
are sufficiently similar to be described together.   These belong
to the same class, and are employed in similar cases to Leuk
and Pisa, being what I have described as earthy springs, the
chief saline ingredient of which is sulphate of lime.   Besides
this they also contain a small quantity of sulphates and
chlorides of soda, potash, and magnesia, together with traces
of iodine, strontium, and arsenic, the latter in imponderable
quantities.   *

CHAPTER XLVIII.

THE SPAS OF ITALY.

FEW countries are so rich in mineral and thermal springs as Italy. Almost all the Italian spas are situated in the valleys at the foot of the mountain chains which intersect that country. Many years since, Dr Gairdner remarked that, " in the prolongation, southwards of the Italian peninsula, all its mineral waters are met with on the Mediterranean side of the Apennines, and none on the Adriatic."* The explanation of this curious fact is to be found, as I have shown in the introductory chapter on mineral and thermal springs, in the position of the volcanoes, and other evidences of subterranean igneous action in that part of Italy.

Commencing our tour through the Italian watering-places in Lombardy, the first of these spas that we meet with is ACQUI, a very ancient town of 9000 inhabitants, in a mountainous district, about thirty miles from Genoa, and a drive of an hour and a half by railway from Alessandria. The mineral springs originate in limestone rock, and are divided into cold and thermal sources. They are mildly sulphurous; but are so slightly charged with chemical ingredients, that, diluted with half their quantity of ordinary water, they are employed by the housewives of Acqui for all domestic and culinary purposes. The warmest source in the town has

* " Essay on Mineral and Thermal Springs," by Meredith Gairdner, M.D., p. 141.

a temperature of 124°; is slightly sulphurous and saline, and is perfectly clear and limpid.

Acqui is seldom resorted to as a sulphurous spa, being inferior in chemical strength to most waters of that class. But the mud-baths of Acqui are remedies of considerable power. The "humus," or mineralised mud, is collected in small chambers, into which the patient enters, and lying down, is completely, with the exception of his head, covered with a thick layer of the "humus," as hot as he can bear it, and remains thus immersed for about three-quarters of an hour, immediately after which a warm bath of the mineral water is administered.

The first effect of the sulphurous mud-bath is to excite a strong determination of blood to the surface, followed by a profuse perspiration, which renders it dangerous for the patient to expose himself to the open air for a considerable time after the bath. The therapeutic influence of this application is most evident in chronic articular enlargements, rheumatic-arthritis, some indolent tumours, intractable cases of secondary syphilis, and rheumatism.

The next spa in our itinerary is ABANO, the birthplace of Livy, which also lies in the same fertile province of Lombardy as Acqui, and is only six miles from Padua. Small as the place is, the bath establishment and the accommodation for visitors are both excellent.

The springs issue from the foot of the Euganean hills, and belong to the class of hot sulphurous saline waters. As at Franzensbad and Acqui, the sulphurous mud of Abano is used for local and general baths. The effects of these mud-baths differs so little from those of others of the same kind that they need not be redescribed.

Tuscany contains several important spas; the first of which that I visited was PISA, where the mineral baths have been in use since the middle of the twelfth century. The thermal sources rise from a calcareous spar rock, at the foot of Mount St Julian, where, within an area of seventy paces, there are no less than twelve springs, which vary in temperature from 106° to 81°. They all belong to the class of thermal salines, and leave a calcareous incrustation in the wells, even in the baths a pellicle of the same character, consisting of salts of lime and magnesia, floats on the surface of the water.

The Pisan mineral springs are used internally in chronic hepatic complaints; in gravel, and some renal affections, in chlorosis, in dysentery and dyspepsia, attended with pain and vomiting. The warm baths are employed in the treatment of gout and rheumatism, impaired power, and enlargement of the joints; also in certain chronic ulcers and skin diseases.

From Pisa, a journey of less than an hour, by the " Strada Ferrata dell' Alta Italia," brings us to LUCCA, five leagues from which are the baths of the same name. They are situated at the foot of Monte Corsena, in one of the prettiest valleys in Tuscany. The watering-place itself consists of a long street of handsome hotels, shops, and lodging-houses, and nothing seems left undone to render this place one of the most agreeable residences in Italy.

There are five or six bathing establishments, one of which is reserved for the poor gratuitously, and the charges of all are extremely moderate. The various sources differ little from each other, except in temperature, in which they vary from 88° to 133°. The principal mineral ingredients of the Lucca springs are the sulphates of magnesia, lime, and aluminia, together with smaller quantities of the carbonates and chlorides of the same bases, also silicate of iron, and traces of iodine and bromine, the total amount of saline matter being about fifteen grains in each pint.

Some of the bath-houses of Lucca were constructed early in the sixteenth century; and ever since that time the soothing and sedative properties of these waters have widely extended their employment in the treatment of cases of chronic rheumatism, leucorrhœa, catarrhal affections of the urinary organs, dyspepsia, and certain cutaneous diseases.

Besides Pisa and Lucca, Tuscany contains many other spas, as for instance, MONTE CATINO between Lucca and Pistoia. The two principal sources are muriated salines, having temperature respectively of 68° and 86°. Both these contain sulphates of lime and aluminia, muriate of soda, and carbonic acid; but in such different proportions, that one, that known as the "Tettuccio," is strongly purgative, while the other, the "Bagnuola," is only slightly aperient. The latter is generally prescribed, and is used chiefly in enlargements of the liver, and in cases of chronic dysentery.

Near Florence are the warm springs of SAN CASCIANO, and in the same department the sulphurous waters of VOLTERRA, with several others. In the old Roman States the only spas I had any opportunity of seeing were the warm salines of CIVITA VECCHIA, which have a temperature of 86°. The Campagna also abounds in hot sulphurous springs, and at VITERBO and PORRETTA are similar sources. The latter of these is the most important. It is strongly sulphurous and very gaseous, the gas consisting principally of carburetted hydrogen, a circumstance of which the guides often take advantage, to startle unscientific visitors, by approaching a torch over the well, on which the gas takes fire, and a luminous vapour floats over the water. Besides their sulphurous constituents, the sources of Porretta contain a great quantity of organic matter, and salts of lime and soda, with traces of iodine. The taste is bituminous and nauseous, and the temperature varies from 86° to 100°. The action of

Porretta water is strongly purgative and diuretic. It is used internally as well as for baths, in the treatment of enlargements and congestions of the abdominal viscera, and in chronic skin diseases.

From the Roman to the Neapolitan spas the transition is natural; and with a brief account of the latter I shall close this chapter and my work.

Some years ago, having had occasion to pass a summer in Naples, I occupied myself in visiting the neighbouring watering-places, the most celebrated of which, CASTELLAMARE, enjoys a high reputation with Neapolitan physicians in the treatment of chronic rheumatism and gout.

Four of the Castellamare sources contain salts of iron, the principal being the "Acqua Ferrata," rising in the Strada Cantieri. Four of the wells are saline, their chief ingre-dients being muriate and sulphate of soda, with chloride of calcium. Four are sulphurous and chalybeate, containing sulphate of iron, with a large volume of sulphuretted hydrogen gas. These, though so distinct in character, all rise within a small area at the base of the hill on which the town stands.

The physiological and therapeutic action of the various mineral springs are, as is indicated by the foregoing classification, necessarily very different. The ferruginous ones, especially the "Acqua Ferrata del Pozzillo," are very powerful chalybeates. The "Acqua Media" is a saline aperient, not unlike Seltzer water in taste.

The action of the suphurous springs of Castellamare need not be here described, as they are similar in effect to others of the same class, and are chiefly used by the local physicians in the treatment of chronic skin diseases, and arthritic affections.

On the opposite side of the Bay of Naples from Castellamare, lies the island of ISCHIA, the next and last of the watering-places which I have to describe. Ischia presents, on every side, the most unmistakable proofs of the volcanic convulsions of which it has been the scene, and on Monte Epomeo twelve separate volcanic cones may still be traced. This mountain, which gives Ischia its peculiar pyramidal aspect, rises in the centre of the island, in a series of highly-cultivated terraces, to the height of between 2000 and 3000 feet.

In summer Ischia is much frequented by the Neapolitans, as a cool maritime residence; but its chief recommendation at all seasons is the reputation of its mineral waters. There are three small towns on the island. Of these Casamicciola is the most resorted to, is prettily situated on a rising ground near the sea, has a population of about 3000 inhabitants, contains some comfortable lodging-houses, and is quite close to the mineral sources.

The Gurgitello, which is the most frequented spring, rises in the Val Ombrasco, near Casamicciola. The temperature of the water is 167°, and its chief ingredients are muriate of soda, carbonate of soda, and sulphates of soda and lime, together with a large amount of free carbonic acid gas. Its medicinal use is almost entirely in the form of baths, in diseases of the periosteum, and in the treatment of sciatica and chronic rheumatism.

The Acqua di Citara rises on a sandy bay, about a mile south of Foria. Its temperature is 120°, and its principal constituents are sulphate and muriate of soda. Its action is chiefly that of a refrigerent and saline laxative, and it is, moreover, used in certain uterine complaints, and in some diseases of the female breast.

The Acqua di Cappone somewhat resembles Wiesbaden water in its similarity of flavour to weak chicken broth. The

temperature is 98°, and it differs from the Gurgitello principally in strength. Its chief use is in dyspepsia and chronic
gastro-intestinal derangements, and also in some uterine
affections.

The Acqua di Bagno Fresco is an alkaline water, said to
possess the property of rendering the skin white and soft, and
therefore specially resorted to by the Neapolitan fair, as well
as by those suffering from certain chronic skin diseases, in
which a stimulating remedy is required.

The " Stufe," or natural vapour baths, of Castiglione, in the
mound of lava close to Lacco, are heated by steam, rising
through crevices, in two small chambers, in the mass of lava.
The temperature of the steam baths is 130°, and they are
employed in the treatment of chronic rheumatism, gout,
and rheumatic-arthritis of long standing. They are also
sometimes prescribed in obstinate skin diseases. As, however, these baths are powerful stimulants and excitants,
they should on no account be used in any case without the
advice of a local physician.

My task ends here. I have now accompanied my reader
through the principal Southern Health Resorts and Foreign
Spas ; and, as I trust, aided the valetudinarian traveller in
pursuit of health, with the counsel of his medical adviser, to
select the climate or mineral water most suitable for his
condition.

INDEX.

272

INDEX.

www.ingramcontent.com/pod-product-compliance
Lightning Source LLC
Chambersburg PA
CBHW020503270326
41926CB00008B/725